DESIGN OF REINFORCED CONCRETE ELEMENTS

SECOND EDITION

P.J.B. Morrell, BSc, PhD, CEng, MICE

Lecturer in Structural Engineering
School of Engineering and Applied Science
University of Sussex

BSP PROFESSIONAL BOOKS

OXFORD LONDON EDINBURGH

BOSTON MELBOURNE

First Edition published by
   Crosby Lockwood Staples 1977
Reissued by Granada Publishing 1982
Reprinted 1984
Second Edition published by
   BSP Professional Books 1989

British Library
Cataloguing in Publication Data
Morrell, P.J.B.
   Design and reinforced concrete elements.
   — 2nd ed.
   1. Reinforced concrete structural
   components. Design
   I. Title
   624.1′8341

ISBN 0−632−02405−4

BSP Professional Books
A division of Blackwell Scientific
   Publications Ltd
Editorial Offices:
Osney Mead, Oxford OX2 0EL
   (Orders: Tel. 0865 240201)
8 John Street, London WC1N 2ES
23 Ainslie Place, Edinburgh EH3 6AJ
3 Cambridge Center, Suite 208, Cambridge,
   MA 02142, USA
107 Barry Street, Carlton, Victoria 3053,
   Australia

Set by Setrite Typesetters Ltd.
Printed and bound in Great Britain by
MacKays of Chatham, PLC, Kent

Contents

Preface

This book is written as an introduction to the design of reinforced concrete structural elements. Its production was undertaken because it was felt that much of the contemporary literature has become too specialist and detailed in nature, and although being very suitable for advanced students, does not offer an easily understood introduction to the subject.

The treatment offered here is therefore to examine the behaviour of reinforced concrete elements from first principles so that a firm understanding of the basic behaviour of reinforced and prestressed concrete can be obtained. Reference is then made to the relevant British Standards to show how these basic principles are embodied into the Standards, but the book makes no attempt to be a British Standards commentary or explanatory document. This introductory approach is aimed at ensuring that students can understand the reasoning behind the major clauses of the Standards, and can appreciate the thinking behind their development rather than adopting a blind adherence to the recommendations. Understanding gained in this way will enable the student to attempt, with confidence, design problems of a non-standard nature, whilst still working within the general guidelines of good and efficient practice.

This second edition has been completely revised to cover changes that have resulted from the replacement of CP 110 with BS 8110. In addition, it has been expanded to give a greater coverage of topics such as moment redistribution and the collapse analysis of slabs, which now includes a more detailed treatment of the strip method of analysis as well as covering the yield line approach. The revision has also been extended to include composite construction, although, since the code of practice relating to composite construction is being rewritten, and as reaction to the preliminary draft edition has been such that publication of the code is still some time away, it has not been possible at the time of writing to know exactly what conclusions and recommendations will be incorporated into the revised code. Equations have therefore been reworked adopting a limit state approach, and although it is recognised that these may not be identical to the final equations produced in the code, it is anticipated that the treatment and final equations will be very similar, so that the student will still be able to follow their development.

ix

Because of its introductory nature, there are no specific references to research reports or learned journals in the text, but additional material is listed at the end of chapters. These suggestions are not intended to be a complete list, but will enable the student who wishes to pursue a particular topic further to obtain details of research in the area, and knowledge of conclusions drawn from that research.

Acknowledgement

Extracts from BS 8110: Part 1: 1985 and Part 3: 1985 are reproduced by permission of the British Standards Institution, Linford Wood, Milton Keynes, MK14 6LE, from whom complete copies can be obtained.

Notation

The notation used throughout is that used in the current British code of practice BS 8110 1985 *The structural use of concrete*. It is appreciated that, where reference is made to earlier codes of practice or to British Standards, some difficulty may be found in identifying equations and recommendations, but it is considered that this will prove less confusing to the reader than if two different sets of notation were used.

A_c Area of concrete
A_{ce} Total equivalent area of concrete
A_{ps} Area of prestressing tendons
A_s Area of tensile reinforcement
A_s' Area of compressive reinforcement
A_{sc} Total area of longitudinal reinforcement
A_{sv} Area of stirrups or links
E_c Modulus of elasticity of concrete
E_s Modulus of elasticity of steel
F_c Longitudinal compressive force
F_{sc} Longitudinal force in compression reinforcement
F_t Longitudinal tensile force
I_c Second moment of area of concrete section
I_{ce} Second moment of area of the equivalent concrete section
I_{se} Second moment of area of the equivalent steel section
M_D Dead load moment
$M_{L_{max}}$ Maximum live load moment
$M_{L_{min}}$ Minimum live load moment
M_{Ls} Live load moment (sagging)
M_U Ultimate moment of resistance, or bending moment due to ultimate loads
$\overline{M}$ Range of live load moments
M_x Moment in the x plane
M_y Moment in the y plane
N Ultimate axial load
P Prestress force

P_e	Effective prestress force
P_k	Characteristic load in tendon
V	Shear force due to ultimate loads
V_c	Ultimate shear resistance of concrete or shear force acting on concrete
V_{co}	Design ultimate shear resistance of uncracked section
V_{cr}	Design ultimate shear resistance of cracked section
V_D	Shear force due to dead loads
V_L	Shear force due to live loads
V_p	Vertical component of tendon force
Z_1	Section modulus to Face 1
Z_2	Section modulus to Face 2
$\overline{Z}$	Range of section modulus
a	Deflection
a_u	Column deflection at Ultimate Limit State
b	Width of section
b_e	Effective breadth
b_f	Breadth of flange
b_{sf}	Breadth of flange ⎫
b_{sw}	Breadth of web ⎬ steel section
b_v	Breadth of section
b_w	Breadth of web or rib
d	Effective depth of section (to tension reinforcement)
d'	Depth to compression reinforcement
d_g	Depth to centroid of steel section
d_n	Depth to centroid of the compression zone
e	Eccentricity
f_{bs}	Local bond stress
f_{bu}	Design ultimate anchorage bond stress
f_c	Compressive stress
$f_{cc_{adm}}$	Allowable compressive stress in concrete
$f_{cc_{adt}}$	Allowable compressive stress in concrete at transfer
$f_{ct_{adm}}$	Allowable tensile stress in concrete
$f_{ct_{adt}}$	Allowable tensile stress in concrete at transfer
f_{cu}	Characteristic concrete strength
f_{pb}	Stress in tendon at failure
f_{pe}	Effective tendon stress
f_{pu}	Characteristic strength of prestress tendons
f_s	Stress in reinforcement
f_t	Tensile stress
f_y	Characteristic strength of reinforcement
f_{yv}	Characteristic strength of shear reinforcement (which should not be taken as greater than 460 N/mm^2)
f_1	Stress at Face 1

f_2	Stress at Face 2
f'_c	Stress in concrete at the level of compression reinforcement
h	Depth
h_f	Flange depth (flange thickness)
k	Constant
l_e	Effective length of column
l_x	Length of shorter side of slab
l_y	Length of longer side of slab
m_x	Moment/length in the x plane
m_y	Moment/length in the y plane
p_{cc}	Permissible stress in concrete in compression
p_{cb}	Permissible stress in concrete in bending compression
p_{sc}	Permissible stress in steel in compression
p_{st}	Permissible stress in steel in tension
s_v	Spacing of stirrups
u	Critical perimeter for punching shear
v	Shear stress
v_c	Design concrete shear stress
v_{co}	Ultimate shear stress in uncracked section
x	Depth to the neutral axis
x_e	Depth to the balanced neutral axis
y	Distance to the extreme fibre
z	Lever arm
α	Constant, loss ratio
α_e	Modular ratio
α_{sx}	Slab bending moment coefficient (x direction)
α_{sy}	Slab bending moment coefficient (y direction)
β	Constant
β_b	Moment redistribution factor
β_{sx}	Slab bending moment coefficient (x direction)
β_{sy}	Slab bending moment coefficient (y direction)
γ_f	Partial safety factor for load
γ_m	Partial safety factor for materials
ϵ	Strain
ϵ_c	Strain in concrete
ϵ_p	Initial strain
ϵ_s	Strain in reinforcement
θ	Rotation
ϕ	Diameter
ω	Load intensity

1 Materials

1.1 Introduction

Concrete is an artificial substance formed by mixing cement, water and aggregate together in such a fashion that when the mixture is allowed to set it forms a stone-like material. The foundation of the compound is the cement, which when mixed with water forms a paste that bonds the aggregate together. Although many materials have the ability to act as cementing agents, for constructional purposes the term 'cement' is used to refer to a lime-based compound. The type most commonly used in Britain is Portland cement, which is formed by mixing naturally occurring substances containing calcium carbonates (chalk and limestone) with substances containing alumina, silica and iron oxide. There are many different forms of Portland cement as well as alternative cement-type materials; comprehensive coverage, together with details of Portland cement manufacture, is given by Neville (1981).

The aggregate in a concrete mixture consists of the non-cementaceous solid particles. These may be composed of many materials, although the most commonly used (for obvious economic reasons) are naturally occurring sands, gravels and crushed rocks. The aggregate is graded so that the finer aggregate occupies any voids that occur between the larger particles, thus giving a denser and stronger concrete. In grading the aggregate, a broad classification of 'fine aggregate' and 'coarse aggregate' is used, the fine aggregate largely comprising sand and dust having a particle size of less than 5 mm. Aggregate above this size is considered to be 'coarse' and is usually further subdivided into groups according to size.

At the time of mixing the constituent elements (including the water) together, the concrete has a consistency which permits it to flow, and therefore needs to be restrained in the position in which it is required. This is achieved by means of moulds or formwork, traditionally of wood or metal, but nowadays also of glass reinforced plastics. As soon as the water is added to the cement, the chemical reaction which sets the mixtures starts, so that the consistency of the concrete mix constantly changes, becoming stiffer and stiffer until it eventually forms a solid. This

1

is achieved within a few hours, but the material continues to gain in strength for many years. The final strength that may be achieved is dependent upon several factors, including the amount of water added at the time of mixing, the degree of compaction and the proportion of aggregate to cement.

1.2   Properties of the concrete mix

When appraising the properties of a concrete mix, the engineer is concerned with two different states of the material: firstly the freshly mixed state, in which the properties of interest are those that govern ease of transport and placing of the mix in the required location, and secondly the hardened state, in which knowledge is needed of the ability of the material to carry the required loads and stresses. The properties of the two states are interrelated, since the final strength of a given concrete mix is greatly affected by the degree of compaction that was achieved when the fresh mix was placed in position. Compaction is the process of driving out air voids that are trapped in the freshly mixed concrete, and is effected by agitating the mix in some way, thus obtaining as dense a mix as possible. The agitation is usually done mechanically, by vibrating either the mould or the concrete mix itself. Vibration by hand is now invariably reserved for small quantities, such as test samples.

The effect of air in concrete is quite pronounced. A fully compacted concrete is taken as being a concrete in which the trapped air is less than 1 per cent by volume. For every 1 per cent above this figure, the compressive strength of the concrete drops by approximately 5·5 per cent. This value is somewhat artificial, however, since the addition of air to a mix permits a lower water: cement ratio to be used, which itself leads to higher-strength concrete; the actual strength loss need not therefore be as high as suggested above. On the other hand, excessive vibration to remove air may lead to segregation of the mix into its constituent elements—a situation that is equally undesirable.

The ease with which compaction may be achieved is largely dependent upon the workability of the mix, which is also a measure of the ease with which fresh concrete may be transported and placed in position. A stiff mix is said to be of low workability, whereas a more fluid mixture has a higher workability. The degree of workability is largely influenced by the amount of water that is added to the mix in its initial stages. Since a more highly workable concrete is easier to place in position, it is also more economical in construction, so that there is a temptation to add water to the mix in excess of the amount required for the chemical reaction to begin. Excess water that is not taken up by the chemical reaction within the cement remains to form voids which weaken the concrete, so that adding extra water to the mix can in fact lead to loss of strength. For given proportions of materials there is a corresponding proportion of water to produce concrete of the greatest strength. Below this (even only 10 per cent below) the chemical reaction may be inhibited to the point where the concrete fails to set.

The amount of water to be added to a given mix is specified in terms of the water : cement ratio, i.e. the weight of water in the concrete divided by the weight of cement. A typical value of the water : cement ratio for a 1 : 2 : 4 mix might be in the order of 0·6; Fig. 1.1 shows the variation in strength that occurs with increasing water : cement ratio for a fully compacted concrete. The theoretical line shows a continuing gain in strength with falling water : cement ratios, even at very low ratios (provided that the minimum amount of water for the chemical reaction has been added). In practice, however, it is found that full compaction is impossible at very low water : cement ratios, so that for practical purposes the strength curve falls away as shown. Figure 1.1 also shows that the fall of strength with increasing water : cement ratio is quite pronounced; it is therefore necessary, when calculating the amount of free water that must be added to the mix, to allow for the water that may be contained within the aggregate.

This brief discussion shows the care that must be taken in adding water to a concrete mix: although addition of water increases the workability, excess water reduces the strength of the concrete and leads to a greater possibility of segregation during compaction.

A measure of the workability of a mix may be obtained from a number of tests, the most common of which are the slump and compaction factor tests which are described in BS 1881: *Methods of testing concrete*. The desired workability is governed by the uses to which the particular mix is to be put, and the difficulty that is likely to be encountered in placing the

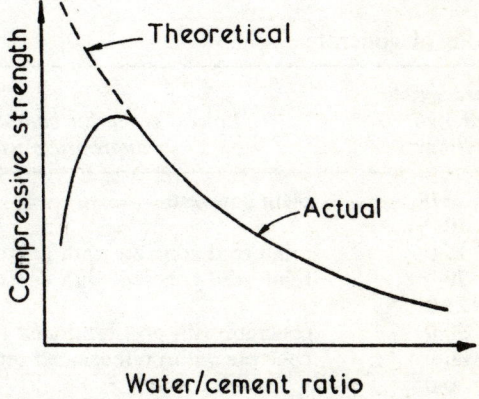

Fig. 1.1 Influence of water : cement ratio on the compressive strength of concrete.

concrete in position: a section in which the reinforcement is highly congested, so that internal vibration is very difficult or even impossible to achieve, requires a concrete having a high workability; whereas for a mass concrete foundation in which virtually no reinforcement is provided, so that problems of compaction are considerably eased, a concrete mix of much lower workability may be used.

Although the properties of the fresh mix are of interest at the constructional stage, it is the properties of the hardened concrete that really concern the designer. These properties can be defined only in general terms, since specific properties vary according to the specification of the concrete. This in itself is not unusual: most materials show some variation in specific properties according to the material composition, but concrete is unusual in that its composition is more readily varied than that of other materials. The main influence on the various properties comes from the actual mix proportions, i.e. the ratios of cement and aggregate. The proportions in which the cement, fine aggregate, and coarse aggregate are combined to provide the concrete mix are traditionally described in terms of the volumetric ratios of the three substances; thus a $1:2:4$ mix has one part of cement to two parts of fine aggregate to four parts of coarse aggregate (all by volume). The less cement that there is in the mix, the weaker is the final concrete (assuming that all other factors influencing the strength remain constant), so that a $1:1:2$ mix is stronger than a $1:2:4$ mix.

Recent advances in concrete technology, however, together with the greater quality control resulting from the now widespread use of central batching plant and subsequent transportation of ready-mixed concrete, have resulted in concrete being specified primarily by its 'grade'. The grade of concrete is an indication of its characteristic strength: Grade C20 concrete has a characteristic strength of 20 N/mm^2, Grade C40 concrete a

Table 1.1   Grades of concrete

Grade	Characteristic strength (N/mm^2)	Lowest grade for compliance with appropriate use
C7	7·0	plain concrete
C10	10·0	
C15	15·0	reinforced concrete with lightweight aggregate
C20	20·0	reinforced concrete with dense aggregate
C25	25·0	
C30	30·0	concrete with post-tensioned tendons
C40	40·0	concrete with pretensioned tendons
C50	50·0	
C60	60·0	

characteristic strength of 40 N/mm^2, etc. Commonly used grades of concrete together with their characteristic strengths and some indication of suggested application are given in Table 1.1.

The durability of the hardened concrete is affected by several factors. Among the most important are the water : cement ratio, workmanship through proper compaction, cover to the reinforcement and the amount of cement in the mix. The minimum cement content varies according to the type of cement, the exposure conditions, and the aggregate sizes, and details are given of this plus the nominal cover that should be provided in Table 3.4 of BS 8110 *The structural use of concrete*.

1.3 Strength of concrete

Discussion of the strength of concrete introduces the primary deficiency of the material of which every engineer is aware: although having great compressive strength, concrete is very weak in tension. It is this lack of tensile strength that leads to the necessity of reinforcement, which carries any tensile forces present in the structure. In general, the behaviour of the material under uniaxial stress only is considered; although it is quite possible to examine its behaviour under other stress conditions, the inaccuracies introduced into the analysis by the assumption of a general uniaxial condition are outweighted by the simplifications in the design process that the assumption permits.

In examining the behaviour of concrete, it is assumed that the proportions of the mix, water content and degree of compaction are constant. Since, at the time of mixing, concrete has zero strength in both tension and compression, it is obvious that a gain in strength takes place with age. The rate of gain in strength depends on the type of cement used; Fig. 1.2 illustrates a typical rate of gain of strength of an ordinary Portland-cement concrete. As may be seen, the rate of gain of strength is rapid in the early life of the concrete, but decreases as the concrete ages, although some cements continue to gain in strength for many years.

For design purposes it is usual to take the concrete strength at 28 days, which is approximately 80 per cent of the strength at one year; any gain in strength after 28 days may be considered as an additional contribution to the factor of safety. If the requirement of the structure is such that the loads will not be applied until later in the life of the structure, e.g. 3 months after the concrete is cast, then the gain in concrete strength between the notional 28-day period and the actual time of loading may be taken into account in the design. In this instance the increase over the 28-day strength would be approximately 16 per cent, varying slightly according to the grade of concrete.

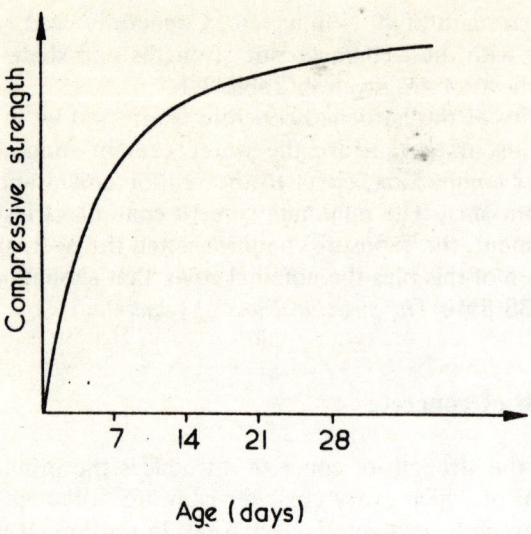

Fig. 1.2 Rate of gain in compressive strength of concrete.

1.3.1 *Stress-strain relationship*

The stress-strain relationship for concrete under uniaxial compression follows the general shape of the curve shown in Fig. 1.3. In the initial stages of loading, the curve is approximately linear, but at a stress

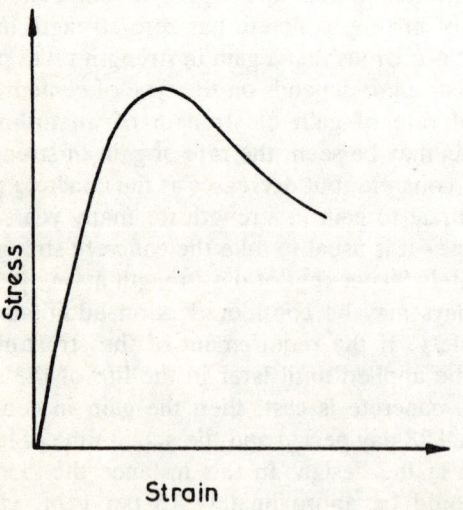

Fig. 1.3 Typical stress-strain curve for concrete under uniaxial compression.

approaching half of the maximum the curve becomes non-linear, with the gradient constantly decreasing as the stress increases. After reaching maximum stress, the slope of the curve reverses to show a 'falling branch' at which time the stress is actually decreasing while the strain continues to increase (this part of the curve is very difficult to obtain). The stress-strain curve may show wide variation, even for concretes of the same specifications, as the rate of loading has considerable effect on the stress-strain relationship. The tensile stress-strain curve is much more uniform, and may be assumed to be linear up to failure, which occurs at a strain of 100–200 microstrain.

The standard test of concrete strength used in Britain is the cube test, in which a concrete cube is loaded in direct compression. Although the load is applied uniaxially, the frictional effect of the loading platens on the end faces of the cube opposes the lateral expansion of the specimen, so that a true uniaxial load is not achieved. The compressive strength of the concrete as indicated by the cube test may exceed the uniaxial strength by up to 25 per cent, and a more accurate indication is obtained by testing a cylinder of concrete which, if sufficiently long in relation to its width, permits the effects of the platen friction to become negligible over the central portion of the cylinder, thus giving a uniaxial stress condition. This condition may be obtained if the cylinder height:diameter ratio is 2:1 or greater.

The tensile strength of concrete is obtained either from a flexural test (also known as the 'modulus of rupture' test) or from an indirect test known as the 'split cylinder' test. The flexural test involves placing an unreinforced concrete beam section under a four-point load configuration so that the middle portion of the beam is under pure bending. The load is increased until the beam fails (by tensile failure in the area of maximum bending moment) and the failure stress is evaluated from simple bending theory, assuming the stress-strain relationship to be linear.

The split cylinder test involves loading a cylinder of concrete (of the same type as may be used for compression tests) in diametrical compression, as shown in Fig. 1.4. By assuming that the material behaves in a linear elastic fashion, it can be shown that an element on the diameter of loading is subjected to a compressive stress along the diameter, and a tensile stress normal to the diameter. If the element is taken to be near the load application point, then the normal tensile stress becomes compressive, and the distribution of horizontal stress across the cylinder diameter follows the form shown in Fig. 1.4(b). Although the horizontal compressive stress in the area near the load is high, a correspondingly high vertical compressive stress produces a state of biaxial compression and ensures that the cylinder fails by tensile splitting rather than by compressive crushing.

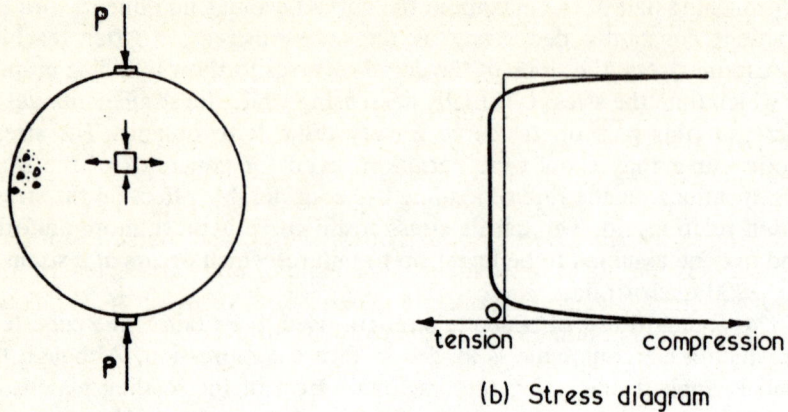

(b) Stress diagram

Fig. 1.4 The split cylinder test, showing the distribution of stresses perpendicular to the load action.

1.3.2 Creep and shrinkage

Even though the stress-strain curve for concrete under compression is substantially linear in the initial stages of loading, determination of the elastic constants of the material is difficult. This difficulty is caused by the creep that occurs in the material when under load, and the shrinkage that is a feature of the setting action. Creep is a measure of the change of strain with time under conditions of constant stress, and can lead to strain increases that are as large as the initial loading strains.

The main effect of creep is to cause change in the value of the modulus of elasticity, which decreases over a period of time. This leads to increased deflection of the structure and also, to a certain extent, to changes in the stresses within the structural materials (an effect that is further considered in Chapter 2). The amount of creep is affected by a number of factors including environmental conditions, composition and strength of the concrete, size of the member, time for which the load has been applied, stress levels in the concrete and type and size of aggregate used. The effects of these various factors on creep are discussed in great detail by Neville (1981) and are recommended for further study.

The overall impact of creep on the Ultimate Limit State is minor, and may be safely ignored although the situation is rather different for serviceability conditions due to the variation of the elastic modulus. Exact allowance for creep effects is both tedious and difficult, so that in practice, creep effects are covered by limiting the span-depth ratios of beams to those given in BS 8110.

Shrinkage is a contraction of the concrete which begins immediately after mixing. It may be considered to take two forms: plastic shrinkage and drying shrinkage.

Plastic shrinkage occurs when the concrete is newly mixed, and results from the condition that the cement and water occupy a greater volume unmixed than does the paste obtained when the two materials are mixed together. The amount of plastic shrinkage depends on the cement content of the mix, greater cement content giving greater shrinkage.

Drying shrinkage is a result of loss of water from the concrete as it sets. Part of this shrinkage movement is reversible, and is affected by the conditions under which the concrete is being kept or used: the shrinkage under 100 per cent humidity is much less than that under drier conditions. Shrinkage is more likely than creep to affect adversely the stresses within the structure, since restrained shrinkage causes tensile stresses to develop.

Like creep, shrinkage is affected by the thickness of the member, environmental conditions, composition of the concrete and other factors, so that accurate assessment of the magnitude of the strain caused is extremely difficult. Advice is given on the calculation of creep and shrinkage strains in part 2 of BS 8110.

1.3.3 Modulus of elasticity

Although the stress-strain relationship in the initial stages of loading appears to be approximately linear, when this portion of the curve is enlarged (Fig. 1.5) it becomes apparent that there is no truly linear part of the curve, so that for all practical purposes evaluation of Young's

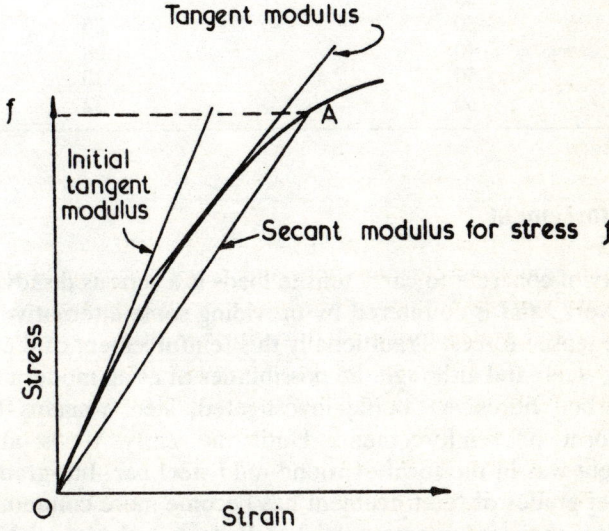

Fig. 1.5 Elastic moduli for concrete under compression.

modulus for the material becomes impossible. It is possible to determine the tangent modulus at any point on the curve, or the initial tangent modulus at the origin, but these apply to such limited areas of the stress-strain curve that they are of little practical value. The shape of the stress-strain curve varies according to the rate of loading: as the rate of loading decreases, so the strain increases. That is because creep takes place over the duration of loading and makes differentiation between elastic strain and creep strain difficult.

Although the total strain varies with the rate of load-application, the variation is small for the range of load-application rates that are usually encountered in laboratory testing. For practical purposes, the deformation during loading is therefore taken as being due to elastic strain only, and any subsequent increase in deformation is taken as due to creep. The elastic modulus obtained by this assumption is known as the 'secant modulus', and is represented by line *OA* in Fig. 1.5. Since the secant modulus decreases with increasing stress, the stress at which the modulus has been determined should always be stated. Typical values of the approximate modulus of elasticity are shown in Table 1.2.

Table 1.2   Values of modulus of elasticity of concrete

Cube strength of concrete at the appropriate age or stage considered (N/mm^2)	Modulus of elasticity of concrete (kN/mm^2)
20	24
25	25
30	26
40	28
50	30
60	32

1.4   Reinforcement

The inability of concrete to carry tensile loads is a serious disadvantage in structural work, and is countered by providing some alternative material to carry the tensile forces. Traditionally this reinforcement of the concrete is of iron or steel, and although the possibilities of using modern materials such as carbon fibres are being investigated, steel remains the most common form of reinforcement. Until the early 1950s almost all reinforcement was in the form of round mild steel bar, but gradually the use of higher grades of reinforcement has become more common, so that today most of the reinforcement used in Britain is of high-yield form.

Most reinforcement is produced by hot rolling processes, and thus has a typical tensile stress-strain relationship, as shown in Fig. 1.6. A marked feature of this relationship is the high strain that takes place after yield and before failure. This is of particular interest to designers of reinforced concrete, and is discussed in Chapter 3. Figure 1.6 shows that an initial straight-line portion of the curve extends from the origin up to the yield stress. (In fact there is some slight deviation from the linear line at a stress somewhat less than yield, but this may be ignored at this stage.) This part of the curve represents the range of elastic behaviour, so that the modulus of elasticity is given by the slope of this straight line. The shape of the stress-strain curve is similar for all steels, and differs only in the value of the yield stress, the modulus of elasticity being for all practical purposes constant. Different qualities of steel are therefore specified in terms of the yield stress.

Some forms of high-yield steel are produced by cold working. This involves loading the bar past its yield point so that it reaches the work-hardening part of the stress-strain curve. If the bar is now unloaded it is found that, on reloading, the curve follows the broken line of Fig. 1.7, and no obvious yield stress or strain plateau is obtained. This means that, for an applied strain greater than the yield strain, the stress in the bar is greater than that which would be obtained in a similar bar that had not been subjected to the cold-working process. Obviously this is of advantage, but the use of such reinforcement has two handicaps. One is that the additional process makes cold-formed worked reinforcement more expensive than hot rolled reinforcement, and the other is that the ductility of the material is reduced by the cold working.

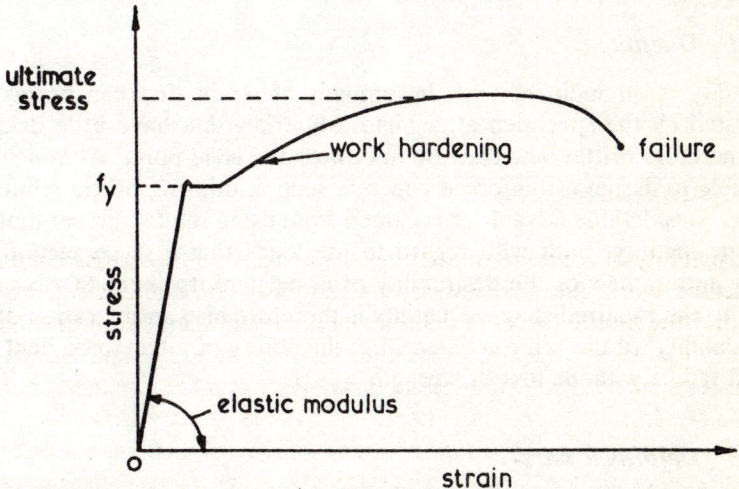

Fig. 1.6 Typical stress-strain curve for reinforcement.

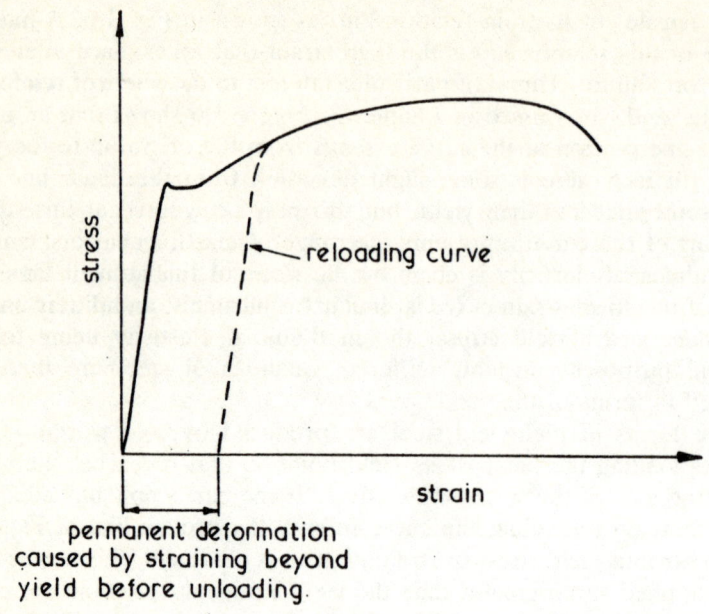

Fig. 1.7 Typical stress-strain curve for cold-worked reinforcement.

The quality of reinforcement that does not have a definite yield point is measured in terms of the proof stress. This denotes the stress which produces a specified percentage of non-proportional elongation or strain, the proof stress normally used being that giving 0·2 per cent elongation.

1.4.1   Ductility

Ductility is an indication of the amount of strain that can be accommodated by the specimen after yield. Materials that have little ductility are therefore brittle, and fracture at or near the yield point. Although it is possible to design a reinforced concrete section utilising brittle reinforcement, considerable advantage is gained from using reinforcement that has a high ductility, both with regard to the loads that a given section can carry and in view of the desirability of using reinforcement that is easily bent to the required shape. Ductility is therefore also an indication of the 'bendability' of the reinforcement, i.e. the ability of a bar to be bent in a small radius without loss of strength.

1.4.2   Deformed bars

A reinforced section is taken as acting homogeneously, the strain in the reinforcement and the strain in the concrete that surrounds it being equal.

This is a result of the bond that develops between the concrete and the reinforcement, which is due to the 'grip' or 'bond' exerted by the concrete in shrinking while setting. The way in which the strain is transmitted from the concrete to the reinforcement via this bond is complex, and it is now considered that the strain (and hence the stress) transfer is carried out in several different stages of behaviour as the strain increases. The 'grip' that the concrete can exert is of limited strength, however, and in cases where high strength reinforcement is used, or where the extent of contact between concrete and reinforcement is too short to enable the bond stress to provide sufficient force, the apparent value of the bond may be increased by providing mechanical keys. This is most commonly done by providing ribs on the reinforcement, or by using a square-section bar that is twisted along its length, instead of a plain round-section bar. Another alternative, discussed in Chapter 8, is to provide hooks in the reinforcement.

Other properties of reinforcement that are of interest to the reinforced concrete designer are weldability (potential loss of strength when welded), fatigue performance, and behaviour under fire conditions. These are all specialist considerations and, as far as reinforced concrete construction is concerned, are covered by ensuring that the reinforcement used complies with the requirements of the appropriate British Standards recommended in BS 8110.

1.4.3 Sizes of reinforcement

For maximum economy of material the reinforcement provided in a structure should correspond with the areas obtained from the design calculations. It is unlikely, however, that this can be achieved without using reinforcing bars of several different sizes, which is confusing in construction and costly in materials. It is invariably more economical to use as few different sizes of reinforcement as possible, even though this may involve using a larger amount of reinforcement than the design actually requires. In general, large bars are cheaper than small bars, so that if all other considerations are equal it is better to use a lesser number of large bars rather than a greater number of small bars. The standard sizes of reinforcing bars, and the lengths in which they are readily available, are discussed in Chapter 8.

In addition to plain and deformed bars, reinforcement may be provided by means of fabric reinforcement. This takes the form of steel wire mesh, usually made from wire having a diameter of between 5 mm and 10 mm, with the mesh size ranging from 100 mm square to 100 mm × 400 mm rectangular. Fabrics are usually used for reinforcing slabs, particularly those that have load and support conditions requiring uniform reinforcement.

Additional reading

NEVILLE, A.M. (1981) *Properties of Concrete*. London: Pitman.
ILLSTON, J.M., DINWOODIE, J.M. & SMITH, A.A. (1979) *Concrete, Timber and Metals*. London: Van Nostrand Reinhold.
JACKSON, N. (Ed) (1983) *Civil Engineering Materials*. London: Macmillan.

2 Elastic or Modular Ratio Method of Design and Analysis

2.1 Introduction

The modular ratio, or elastic method of design has now been largely superseded by methods based on the ultimate load approach, and the introduction in 1972 of CP 110, relating to the structural use of concrete based on the limit state philosophy, furthered the demise of methods based on elastic behaviour. However, the elastic approach still provides the basis for determining certain serviceability conditions, in particular, deflection and is of considerable use in the design of prestressed sections, where, as is shown in Chapter 5, the serviceability condition is often more important than is the case in reinforced sections.

Some confusion is often caused by the use of the terms 'modular ratio', 'elastic design' and 'permissible stress design' when describing design methods for reinforced concrete. In fact these three titles all refer to the same fundamental method, which has as its basis the assumption that the stress-strain behaviour of both concrete and steel is linear elastic. This means that the materials each have a constant modulus of elasticity, so that there is a fixed ratio between the moduli. This is termed the 'modular ratio', and is used to determine the relative magnitudes of stress in concrete and reinforcement. The actual stresses are limited to certain permissible values in order to justify the assumption of the linear elastic stress-strain relationship.

The main criticism of the modular ratio method of design is the assumption of elastic behaviour of the materials. For steel reinforcement, in which the maximum stress is restricted to less than yield stress, this assumption is reasonably accurate; but it oversimplifies the behaviour of concrete under compression, for which a linear stress-strain relationship may only reasonably be assumed provided that the stress does not exceed about half of the ultimate compressive stress, and which even in this low stress region does not truly exhibit either elastic or linear behaviour due to the effects of creep. The assumption does, however, permit a simple design method to be developed so that for design purposes a factor of one-third is applied to the ultimate cube stress to obtain the permissible

15

stress for concrete under flexure, thus ensuring that the concrete behaviour approximates to the assumed linear elastic state. This gives the
impression that a factor of safety of 3 is applied to the concrete stresses.
Since the failure stress for a cube is appreciably greater than the uniaxial
failure stress (see Chapter 1), a factor of one-third applied to the cube
stress reduces to an effective factor of approximately one-half on the
ultimate uniaxial stress, so that the actual factor of safety is rather less
than would at first appear.

The permissible stress in steel reinforcement is related to the yield
stress by a factor of 1·8. The strain in the reinforcement at this stress is
several times that which might be considered as being the ultimate tensile
strain of the concrete surrounding the reinforcement, so that development
of the full permissible stress in the reinforcement implies that cracking of
the concrete in the tension zone must occur. Design calculations accept,
and in fact assume, that tensile cracking will take place, but it is desirable
to limit the width of the crack in order to restrict corrosion of the
reinforcement and to avoid the presence of unsightly, and to the layman
apparently dangerous, effects. The width of the crack is related to the
stress in the reinforcement, so that, in order to limit the crack width, the
permissible tensile stress in steel reinforcement is limited to a maximum
value of 230 N/mm^2 no matter what the yield strength of the steel may be.

The permissible stresses are frequently represented by the letter p, with
suffixes being used to indicate the material and the force producing the
stress. Thus p_{cb} represents the permissible stress in concrete due to
bending and p_{st} the permissible stress in steel in tension.

2.2   Design of a singly reinforced beam

Consider a simple rectangular reinforced concrete beam of breadth b,
depth h and carrying a moment M such that the upper surface of the
section is in compression (Fig. 2.1). The reinforcement is at a distance d
from the top of the section. Thus the cover to the centre of the reinforcement is $(h - d)$. The neutral axis is taken as being a distance x from
the top of the section. The concrete above the neutral axis is under
compression, while the concrete below the neutral axis is assumed to have
cracked under tension and is therefore unstressed. The assumed stress
diagram is shown in Fig. 2.1(b), with the stress in the concrete increasing
linearly from zero at the neutral axis to a maximum at the extreme
compression fibre, and the stress in the reinforcement being taken as
uniform throughout.

The longitudinal forces in the two materials are represented by F_c and
F_t and, since force may be equated to stress × area,

$$F_c = \frac{xbf_c}{2} \quad \text{and} \quad F_t = A_s f_s$$

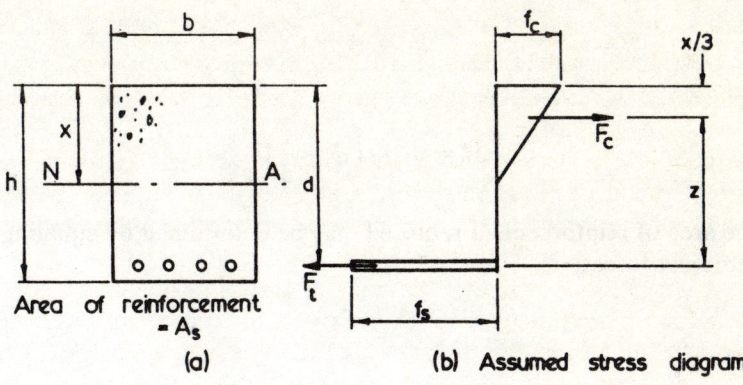

Fig. 2.1 Singly reinforced rectangular beam under flexure.

The action of the forces F_c and F_t is to produce an internal couple which resists the applied bending moment. This couple is known as the 'moment of resistance' of the section, and its magnitude may be determined by taking moments of the internal forces about any point. The most convenient point is usually the line of action of one of the internal forces, since by taking moments about such a point only one of the internal forces appears in the equation. The line of action of the force associated with a particular stress diagram lies through the centroid of that diagram. The compressive force F_c therefore acts through a point $x/3$ down from the top surface of the section, and the tensile force F_t acts through the centre of the reinforcement, i.e. at d from the top surface. If the internal couple is obtained by taking moments about either of the internal forces, then the distance between the lines of action of each of the internal forces represents the lever arm of the couple. This is signified by z.

Taking moments about the line of action of the tensile force, therefore:

$$M = F_c z$$

but

$$F_c = \frac{bxf_c}{2} \quad \text{and} \quad z = d - \frac{x}{3}$$

therefore

$$M = \frac{f_c bx}{2} \left(d - \frac{x}{3} \right) \qquad (2.1)$$

Alternatively,

$$M = F_t z$$

but

$$F_t = A_s f_s$$

therefore

$$M = A_s f_s \left(d - \frac{x}{3} \right) \qquad (2.2)$$

The area of reinforcement required may be determined by equating the longitudinal forces:

$$F_c = F_t$$

therefore

$$A_s = \frac{bxf_c}{2f_s} \qquad (2.3)$$

These equations permit the calculation of the dimensions of the section, together with the amount of reinforcement required for a particular bending moment. The equations are all dependent upon the position of the neutral axis, however, which at this stage is unknown.

One of the basic assumptions made in reinforced concrete design is that the plane sections remain plane. The implication of this is that the strain diagram is linear, so that strain is proportional to distance from the neutral axis (the neutral axis being the level of zero strain). Taking the strain diagram as that shown in Fig. 2.2, therefore, the maximum strain in the concrete will occur at the extreme fibre, distance x from the neutral

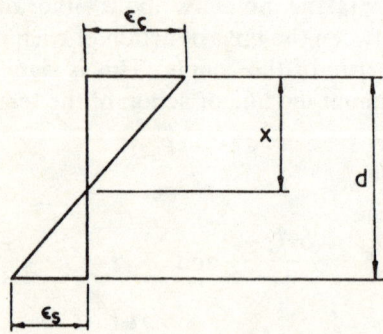

Fig. 2.2 Longitudinal strain diagram for a beam under flexure.

axis, and the maximum strain in the reinforcement will be at a distance $(d - x)$ from the neutral axis.

From similar triangles,

$$\frac{\epsilon_c}{\epsilon_s} = \frac{x}{d - x}$$

$$\epsilon_c = \frac{f_c}{E_c} \quad \text{and} \quad \epsilon_s = \frac{f_s}{E_s}$$

therefore

$$\frac{\epsilon_c}{\epsilon_s} = \frac{f_c E_s}{f_s E_c} = \frac{x}{d - x}$$

By definition, E_s/E_c = the modular ratio α_e, therefore

$$\frac{f_c \alpha_e}{f_s} = \frac{x}{d - x} \tag{2.4}$$

giving

$$x = \frac{f_c \times d}{(f_s/\alpha_e) + f_c} \tag{2.5}$$

so that if the stresses in the reinforcement and in the concrete are known, the position of the neutral axis may be found.

2.2.1 Transformed sections

An alternative method of developing eqs. (2.1)–(2.5) is that known as the method of transformed sections. This involves replacing the reinforcement (for purposes of analysis only) by an equivalent area of concrete. To do this it is necessary to imagine that concrete is able to carry tensile forces; in this case, to obtain a force in the 'equivalent' concrete equal to that in the reinforcement, an area of concrete, A_c, must be used, such that

$$f_c A_c = f_s A_s$$

Substituting for f_c and f_s,

$$\epsilon_c E_c A_c = \epsilon_s E_s A_s$$

therefore

$$A_c = \frac{\epsilon_s}{\epsilon_c} \alpha_e A_s \tag{2.6}$$

However, since the equivalent concrete area is to be placed at the same level as the reinforcement, the strain in the equivalent concrete must equal the strain in the reinforcement. Therefore

$$A_c = \alpha_e A_s \tag{2.7}$$

This forms the basis of the method of transformed or equivalent sections: an area of reinforcement A_s may be replaced by an equivalent area of concrete $\alpha_e A_s$.

The imagined assumption that the concrete is now able to carry tensile stresses applies only to that area of concrete that is considered to be replacing the reinforcement; it does not apply to the concrete that was in the tension zone of the section, and which is still assumed to have cracked and therefore to be unable to carry any tensile stresses.

The stresses in the transformed section are determined from the simple theory of bending, but therefore this theory can be applied the position of the neutral axis must be found. For the flexural condition, the neutral axis may be considered as passing through the centroid of the section. In this case, the section under consideration is the transformed section, so that in order to determine the neutral axis position it is only necessary to determine the centroid of the section, by equating the first moments of area of the transformed section above and below the neutral axis.

Working in equivalent concrete units, therefore, and referring to Fig. 2.1,

$$\frac{bx^2}{2} = \alpha_e A_s(d - x) \tag{2.8}$$

from which x may be obtained. From the simple theory of bending,

$$\frac{M}{I_c} = \frac{f}{y}$$

where I_c is the second moment of area of the section about an axis through the centroid (i.e. about the neutral axis). For the transformed, or equivalent, concrete section,

$$I_{ce} \simeq \frac{bx^3}{12} + \frac{bx}{4}x^2 + \alpha_e A_s(d - x)^2$$

$$= \frac{bx^2}{6}(3d - x) \tag{2.9}$$

so that rewriting the simple bending equation in terms of the concrete stress f_c gives

$$M = \frac{f_c I_{ce}}{y}$$

$$= f_c \frac{bx}{2}\left(d - \frac{x}{3}\right) \tag{2.1 bis}$$

To obtain the reinforcement stress, the second moment of area of the section must be rewritten in terms of the equivalent *reinforcement* section; therefore

$$I_{se} = \frac{I_{ce}}{\alpha_e}$$

and hence

$$M = \frac{f_s I_{se}}{d - x}$$

$$= \frac{f_s b x^2 (3d - x)}{6\alpha_e (d - x)} \tag{2.10}$$

Dividing eq. (2.1) by eq. (2.10) gives

$$\frac{f_c \alpha_e}{f_s} = \frac{x}{d - x} \tag{2.4 bis}$$

From eq. (2.8)

$$A_s = \frac{b x^2}{2\alpha_e (d - x)} = \frac{f_c b x}{2 f_s} \tag{2.3 bis}$$

and substituting from eq. (2.3) into eq. (2.10) gives

$$M = \frac{f_s A_s}{3}(3d - x) \tag{2.2 bis}$$

2.2.2 Balanced sections

The most economical design in terms of material usage is one in which the maximum stresses in both the reinforcement and the concrete reach the permissible stress values. This is known as the 'economic' or 'balanced' section. In practice, however, most sections are uneconomic or non-balanced, since the dimensions and reinforcement areas must be rounded off to suitable practical values. Non-balanced sections are of two types:

(a) Over-reinforced sections
(b) Under-reinforced sections.

Over-reinforced sections are those that contain more reinforcement than is required to give a balanced action. Hence, as the applied moment is increased, the concrete reaches its permissible stress value first, and by the time the reinforcement reaches its permissible stress, the concrete is overstressed. The maximum moment of resistance of an over-reinforced section must therefore be calculated by taking moments of the compressive force about the line of action of the tensile force, i.e. by using eq. (2.1).

Since to continue to increase the load produces overstress in the concrete earlier than in the reinforcement, any failure is sudden, as the concrete crushes in compression.

Under-reinforced sections contain too little reinforcement to permit the balanced action to be developed. In such sections the tensile reinforcement is insufficient to develop the full strength of the concrete in compression, so that when the reinforcement is fully stressed the concrete is understressed. The maximum moment of resistance of under-reinforced sections must therefore be calculated by taking moments of the tensile forces about the line of action of the compressive forces (eq. (2.2)). Failure is more gradual than with over-reinforced sections as, when overstressed, the reinforcement yields but is still able to support the yield stress.

Stress diagrams in equivalent concrete units for both over- and under-reinforced sections are shown in Fig. 2.3. It may be seen that the over-reinforced section is characterised by having a depth to the neutral axis

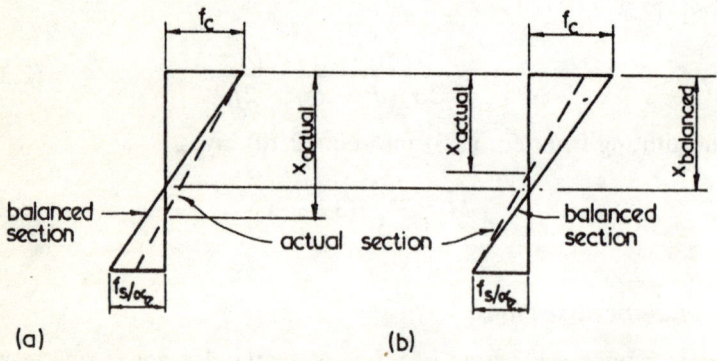

Fig. 2.3   Stress diagrams (equivalent concrete units) for (a) over-reinforced and (b) under-reinforced sections in flexure.

greater than that of the balanced section, while for under-reinforced sections the depth to the neutral axis is less than the balanced depth. In order to determine whether a section is over- or under-reinforced, it is therefore necessary to calculate the positions of both the balanced and the actual neutral axes. The balanced neutral axis is determined from eq. (2.5) by substituting the permissible stresses p_{cb} and p_{st} for the terms f_c and f_s; the actual position of the neutral axis in a rectangular section may be obtained from eqs. (2.3) and (2.4). From eq. (2.3),

$$\frac{f_c}{f_s} = \frac{2A_s}{bx}$$

From eq. (2.4),

$$\frac{f_c}{f_s} = \frac{x}{\alpha_e(d - x)}$$

therefore

$$\frac{2A_s}{bx} = \frac{x}{\alpha_e(d - x)} \qquad\qquad (2.8\ bis)$$

giving a quadratic in x which may be readily solved for a given section.

This alternative method of obtaining the position of the neutral axis leads to the same equation obtained in Section 2.2.1.

For non-rectangular sections it is usually easier to use the transformed section technique to determine the positions of both the actual neutral axis position and the balanced position.

Example 2.1

Determine the permissible bending moment that may be carried by a rectangular reinforced concrete section 0.5 m wide, effective depth 0.7 m, reinforced by four 25 mm diameter bars ($A_s = 1964$ mm^2), if $p_{cb} = 8.5$ N/mm^2, $p_{st} = 210$ N/mm^2 and $\alpha_e = 15$.

From eq. (2.5), determine the balanced position of the neutral axis:

$$x_e = \frac{8.5}{(210/15) + 8.5}\,0.7 = 0.264 \text{ m}$$

The actual position of the neutral axis may be determined from eq. (2.8):

$$\frac{500x^2}{2} = 15 \times 1964(700 - x)$$

therefore

$$x = 0.234 \text{ m}$$

Since the actual value of x is less than the balanced value, Fig. 2.3 shows that the section is under-reinforced, so that the limiting stresses will occur in the reinforcement. Hence the maximum permissible moment is determined from eq. (2.2).

$$M = 1964 \times 210 \times \left(700 - \frac{234}{3}\right) \text{N mm}$$

$$= 256 \text{ kN m}$$

To illustrate the importance of determining whether the section is under- or over-reinforced, consider the same section, but now assume that the section is balanced. As before, x_e is obtained from eq. (2.5):

$$x_e = 0.264 \text{ m}$$

From eq. (2.1),

$$M = \frac{8 \cdot 5 \times 500 \times 264}{2} \left(700 - \frac{264}{3}\right) \text{ N mm}$$

$$= 343 \text{ kN m}$$

or, from eq. (2.2),

$$M = 210 \times 1964 \left(700 - \frac{264}{3}\right) \text{ N mm}$$

$$= 252 \text{ kN m}$$

The importance of considering the actual position of the neutral axis is therefore clearly established, for, had the permissible moment been calculated according to eq. (2.1), a situation of overload by 87 kN (or 34 per cent) could have developed.

2.3 Doubly reinforced sections

It has so far been assumed that the longitudinal reinforcement is restricted to the tension zone only. Frequently, however, reinforcement will be placed in both the tensile and compressive zones. This may be because the physical dimensions of a section are restricted (usually with regard to beam depth, which may be limited through headroom considerations), so that the moment that the beam is required to carry is greater than the balanced moment obtained by the method described above, or because the provision of additional reinforcement is desirable to provide adequate anchorage for shear reinforcement (see Chapter 4) or to accommodate reversal of moment. Some increase in the moment of resistance of the section may be obtained by increasing the area of the reinforcement. This produces an over-reinforced section which has a greater moment capacity than the balanced section, but the increase in reinforcement area that is required to produce a substantial increase in the moment capacity is such that, except for small increases of moment, this method is uneconomical.

A more economical method is to provide additional tension reinforcement plus some reinforcement in the compression zone of the section, thus providing an internal reinforcement couple. The compression reinforcement is usually used uneconomically since its strain must be the same as the strain in the concrete that surrounds it, thus reducing the stress in the compression reinforcement to α_e times the concrete stress at that level. Despite this inefficient usage of the compression reinforcement, however, the method leads to a more economical design than is obtained by providing a grossly over-reinforced section.

2.3.1 Design of doubly reinforced sections

Suppose that the section is required to have a moment of resistance M. The economic moment of resistance M_e is determined in the same way as if the section were a balanced singly reinforced section, using eq. (2.1). This leaves an excess moment $(M - M_e)$ which must be carried by the couple produced by the compression reinforcement and the extra tensile reinforcement. The extra area of reinforcement that must be provided in the tension zone is easily calculated by taking moments about the line of action of the compression reinforcement. Thus,

$$A_{s_{\text{extra}}} f_s z' = M - M_e \tag{2.11}$$

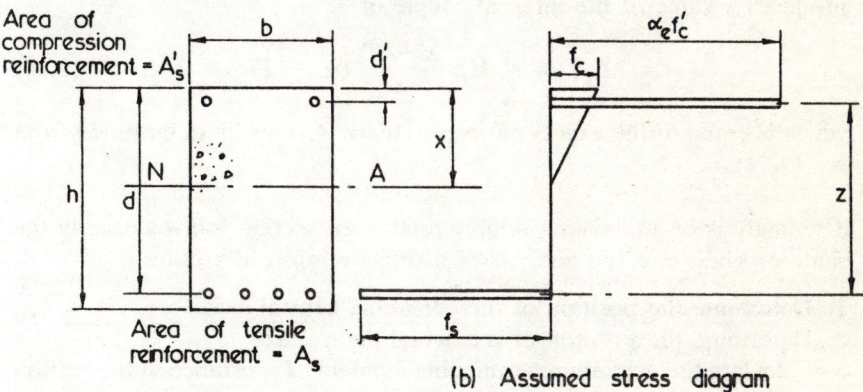

Fig. 2.4 Doubly reinforced rectangular beam under flexure.

where z' is the lever arm of the internal couple produced by the tension and compression reinforcements, and, with regard to design, f_s is the permissible stress of the reinforcement in tension p_{st}.

The stress in the compression reinforcement is obtained from compatibility of the strain in the concrete with the strain in the reinforcement at the same level, so that $\epsilon_c = \epsilon_s$. However, since $\epsilon = f/E$, so $f_s = (f_c E_s)/E_c$; hence $f_s = \alpha_e f_c$, making the stress in the compression reinforcement equal to α_e times the stress in the concrete at that level.

From similar triangles (Fig. 2.4(b)), the stress in the concrete at the level of the compression reinforcement is f'_c, where

$$f'_c = \frac{x - d'}{x} f_c$$

The stress in the compression reinforcement, f_{sc}, therefore equals

$[(x - d')/x]f_c\alpha_e$ and the force in the compression reinforcement is F_{sc}, where

$$F_{sc} = A'_s f_{sc} = A'_s f_c \alpha_e \frac{x - d'}{x} \qquad (2.12)$$

However, there is a loss of concrete area of magnitude A'_s in the compression zone due to the presence of the reinforcement, which causes a corresponding loss in the compression force of $A'_s f'_c$, so that the net gain in compression force is F'_c, where

$$F'_c = F_{sc} - A'_s f'_c = A'_s f_c \frac{x - d'}{x} (\alpha_e - 1) \qquad (2.13)$$

Taking moments about the line of action of the tensile reinforcement produces a value of the internal couple of

$$F'_c z' = z' A'_s f_c \frac{x - d'}{x} (\alpha_e - 1) \qquad (2.14)$$

which is equal to the excess moment. Hence A'_s may be determined from eq. (2.14).

The analysis of an existing doubly reinforced section follows exactly the same procedure as the analysis of a singly reinforced section.

1. Determine the position of the economic neutral axis.
2. Determine the position of the actual neutral axis.
3. Calculate the maximum permissible moment of resistance of the section assuming it to be singly reinforced, and obtain the area of tension reinforcement required to develop this moment.
4. Calculate the excess tension reinforcement in the section.
5. Calculate the maximum permissible excess moment that will be produced by the internal couple. Note that Step 5 requires calculation for both the compressive and the excess tensile reinforcement.

If it is known that the section has been designed according to the procedure detailed earlier, Step 2 of the analysis may be omitted. However, it is not safe to assume that the designer has followed this method of design, and so, in general, Step 2 should not be bypassed.

Example 2.2
Design a rectangular reinforced concrete beam 500 mm wide and 750 mm deep to carry a bending moment of 500 kN m. Take $p_{cb} = 8.5$ N/mm^2, $p_{st} = 210$ N/mm^2 and $\alpha_e = 15$. Allow a cover of 50 mm to the centres of both the tension and compression reinforcement.

Following the procedure detailed above, the first step is to determine the economic moment that the section will be able to carry. From eq. (2.5),

$$x_e = \frac{8 \cdot 5}{(210/15) + 8 \cdot 5} \times 700 = 264 \text{ mm}$$

From eq. (2.1),

$$M_e = \frac{500 \times 8 \cdot 5 \times 264}{2} \left(700 - \frac{264}{3}\right) \text{ N mm}$$

$$= 343 \text{ kN m}$$

The reinforcement required to carry this economic moment is obtained from eq. (2.3):

$$A_s = \frac{500 \times 264 \times 8 \cdot 5}{2 \times 210} = 2671 \text{ mm}^2$$

The excess moment $= M - M_e = 500 - 343 = 157 \text{ kN m}$, which must be resisted by the internal reinforcement couple. From eq. (2.11),

$$A_{s_{extra}} = \frac{M - M_e}{f_s z'} \quad \text{where } z' = d - d'$$

therefore

$$A_{s_{extra}} = \frac{157 \times 10^6}{210 \times 650} = 1150 \text{ mm}^2$$

From eq. (2.14),

$$A_s' = \frac{(M - M_e)x}{z' f_c (x - d')(\alpha_e - 1)}$$

$$= \frac{157 \times 10^6 \times 264}{650 \times 8 \cdot 5 \times 214 \times 14} = 2503 \text{ mm}^2$$

Therefore the total tension reinforcement required is $1150 + 2671 = 3821 \text{ mm}^2$, and the compression reinforcement required is 2503 mm^2.

It is interesting to re-examine the problem assuming the section to be singly reinforced. The economic moment is the same as that obtained above, so that, in order to increase the moment of resistance, the section must be grossly over-reinforced. For an over-reinforced section, eq. (2.1) applies.

The procedure is to use eq. (2.1) to determine the required position of the neutral axis in order to achieve a moment of resistance of 500 kN m, and then to use eq. (2.8) to find the amount of reinforcement necessary to give the neutral axis depth calculated from (2.1).

From eq. (2.1),

$$500 \times 10^6 = 8 \cdot 5 \times 500 \times \frac{x}{2} \left(700 - \frac{x}{3}\right)$$

therefore

$$x = 420 \text{ mm}$$

From eq. (2.8),

$$A_s = \frac{500 \times 420^2}{2 \times 15 \times 280} \text{ mm}^2$$

therefore

$$A_s = 10\,500 \text{ mm}^2$$

Hence, if a singly reinforced section is provided, the total reinforcement area is $10\,500 \text{ mm}^2$, whereas for the doubly reinforced section the total area required is 6324 mm^2, a saving of approximately 40 per cent.

Example 2.3
Having designed the beam of Example 2.2, the designer specifies that the required reinforcement area be obtained by eight 25 mm diameter bars in tension ($A_s = 3927 \text{ mm}^2$), and five 25 mm diameter bars in compression ($A'_s = 2454 \text{ mm}^2$). Owing to some confusion on the site, however, the compression reinforcement actually provided is five 20 mm diameter bars (area $= 1571 \text{ mm}^2$). What effect will this have on the load capacity of the beam?

The problem is now one of analysis. Following the analysis procedure detailed on page 26, from eq. (2.5) $x_e = 264$ mm. By the method of transformed sections,

$$1571 \times 14 \times (x - 50) + 500\,\frac{x^2}{2} = 3927 \times 15 \times (700 - x)$$

therefore

$$x = 280 \text{ mm}$$

i.e. the section is over-reinforced in tension, and the limiting stress will therefore be reached in the compression zone.

From eq. (2.1),

$$M = \frac{8 \cdot 5 \times 280 \times 500}{2} \left(700 - \frac{280}{3}\right) = 361 \text{ kN m}$$

From eq. (2.2), the area of reinforcement required for this moment is $M/(p_{st}z) = 2834 \text{ mm}^2$, leaving an excess area of tension reinforcement of 1093 mm^2.

Taking moments about the line of action of the compression reinforcement,

$$M_e = 1093 \times 210 \times 650 = 149 \text{ kN m}$$

and about the line of action of the tensile reinforcement (eq. (2.14)),

$$M_e = 14 \times 1571 \times 8.5 \times \frac{280 - 50}{280} \times 650 = 100 \text{ kN m}$$

so that the total load capacity of the beam is $100 + 361 = 461$ kN m.

2.4   T and L beams

A common form of construction is that using the beam and slab principle, in which the slab is supported by a system of beams. If the connection between the beam and the slab is able to transmit longitudinal shear force, then the beam and slab may be considered to act as a homogeneous section of T or L form. For beams loaded so that the top surface is in compression, the slab therefore becomes the compression flange of the beam, resulting in a greater zone of compression and giving a more economical section. For a reinforced concrete beam-slab section, adequate connection between the beam and the slab is easily provided by casting the section as a monolithic whole, or by suitable treatment to the surface of the beam before casting the slab on top. For beams other than concrete, some mechanical system is frequently needed to ensure that the longitudinal shear forces do not cause separation of the two components of the section. This is discussed in more detail in Chapter 6.

In order to analyse or design a flanged section it is necessary to know the width of the flange. For an isolated section this is clearly defined, but for a flanged section which forms part of a beam-slab construction, the width of the flange requires assessment. For any flanged beam loaded in flexure, the compressive stress in the upper flange decreases as the distance from the web increases. This is particularly so for flanges that are thin in relation to their width, so that in order to simplify the design procedure it is usual to assume a uniform distribution of stress over a reduced width of flange. This reduced width is known as the 'effective width'.

The effective width of the compression flange of a flanged beam is not constant along the length of the beam, but depends on a number of factors. These include:

(i) The end conditions of the beam
(ii) The method of load application
(iii) The ratio between the flange thickness and the beam depth
(iv) The ratios of the length of the beam between the points of zero moment to (a) the width of the rib (web) and (b) the distance between the ribs of adjacent beams.

The effects of these factors have been summarised in the British codes of practice, which make recommendations allowing the effective width to be readily calculated. For the modular ratio approach, the code specifies that the effective width of a T beam shall not exceed the least of:

 (i) one-third of the effective span of the T beam
 (ii) the distance between the centres of the ribs of the T beams
(iii) the rib breadth plus 12 times the slab thickness.

2.4.1 Design and analysis of flanged sections

The design of flanged beams usually involves a combination of informed assumption verified by analysis. The steps of design are set out below, and although the example is for a T beam, the procedure applies equally to L beams.

1. Assume an overall depth of section. This may be decided by considerations of headroom or other factors, but even if no restrictions to size apply, the depth is influenced by the load that the section is to carry. Typical values range from span/20 upwards.
2. Assume a rib breadth. This will have little effect on the overall strength of the section, but will influence the self-weight and the ease with which the reinforcement may be placed. It is uneconomical to provide too great a rib breadth, and if the rib breadth is too small problems of lateral stability may arise, so that reasonable values are in the range of one-third to one-half of the section depth.
3. Calculate the lever arm. For accurate calculation the position of the neutral axis must be known, but an approximate value can be obtained from $z = d - (h_f/2)$.
4. Calculate the area of reinforcement required from $M = A_s p_{st} z$.
5. Analyse the section obtained from Steps 1–4.

In choosing a section size by the above procedure, no provision is made for determining the effective breadth or the depth of the flange. This is because these dimensions are usually decided by other factors in the overall design, the one by the recommendations of the relevant code of practice, and the other by the design of the slab which forms the flange.

The general method of analysis is exactly the same as that discussed previously for rectangular sections. The position of the neutral axis is determined and compared with the 'balanced' position, thus establishing whether the section is over- or under-reinforced, and allowing the position of the limiting stresses in the section to be determined. The permissible moment of resistance is then calculated from the appropriate equation and compared with the required moment of resistance.

The exact analysis of a flanged section varies slightly according to the position of the neutral axis, which may lie either in the flange or in the rib of the section. If the neutral axis lies in the flange the analysis is identical to that for a rectangular beam having a breadth equal to the effective width of the flange, and the previously derived expressions and equations apply. Where the neutral axis lies in the rib or web of the section, however, the calculation is frequently simplified by ignoring the compressive stress in the area of the rib above the neutral axis. The justification for this is that, since the stress acting on this area is low in comparison with the maximum compressive stress, and as the area on which it acts is also small, the compressive force acting on the area of the rib above the neutral axis is negligible when compared with the overall compressive force. Calculation is therefore simplified by taking the compressive force as being the mean compressive stress on the flange multiplied by the flange area. To illustrate this point, Example 2.4 considers the effect of including the compressive stress on the rib, and shows that the above simplification is in fact perfectly acceptable.

Example 2.4
Determine the moment of resistance of the section shown in Fig. 2.5. Assume $p_{cb} = 7$ N/mm^2, $p_{st} = 140$ N/mm^2 and $\alpha_e = 15$.
    As an initial assumption, assume that the neutral axis lies in the web.

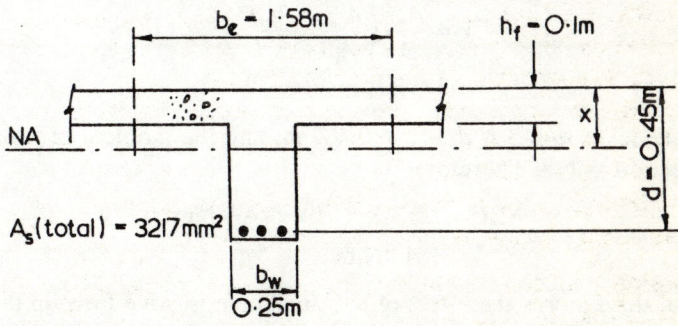

Fig. 2.5

To determine the neutral axis position, consider the first moment of area in equivalent concrete units about the assumed NA:

$$b_e h_f \left(x - \frac{h_f}{2}\right) + \frac{b_w}{2}(x - h_f)^2 = (d - x)A_s \alpha_e$$

Substitution into this equation for the dimensions of the section enables x to be determined:

$$x = 0.1425 \text{ m}$$

i.e. the assumption that the neutral axis is in the web is correct.

The economic or balanced neutral axis position is given by eq. (2.5):

$$x_e = \frac{7}{(140/15) + 7} \, 0.45 = 0.193 \text{ m}$$

Therefore, since the actual neutral axis depth is less than the balanced neutral axis depth, the section is under-reinforced and the limiting stress condition will be reached in the reinforcement.

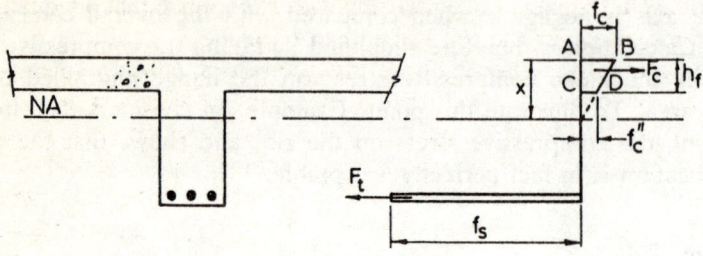

Fig. 2.6

The compressive force F_c is taken as acting through the centroid of the trapezoidal area $ABCD$, distance $\bar{x}$ from the top of the flange, where

$$\bar{x} = \frac{AC}{3} \left(\frac{AB + 2CD}{AB + CD} \right)$$
$$= 0.041 \text{ m}$$

so that the lever arm $z = d - \bar{x} = 0.409$ m, and the moment of resistance of the section $= F_t z$. Therefore

$$\text{MR} = 3217 \times 140 \times 0.409 \text{ Nm}$$
$$= 184 \text{ kN m}$$

Note that this ignores the effect of the small compressive force in the area of the web above the neutral axis. If this is included it will be found that the lever arm reduces to 0·408 m, and the moment of resistance of the section falls by 0·6 kN m—an insignificant reduction. The effect is always much less in under-reinforced than the over-reinforced sections, but in the latter case the simplified approach leads to slight over-design and is therefore acceptable.

2.5 Self-weight

The work so far discussed in this chapter has been concerned either with designing a beam to carry a certain flexural moment, or with analysing an existing section to determine the bending moment which can be supported. It is obviously outside the scope of this book to discuss the methods by which the required moments are obtained, but it is vital to the success of the design that the designer remembers that one component of the bending moment carried by a member is its self-weight. As the span of a beam increases, so the effects of the self-weight of the section become proportionally higher, and estimating the allowance necessary for self-weight becomes more difficult. It is this estimation of the self-weight in the initial stages of the design that frequently causes the most problems to the inexperienced designer, since it is only with experience that adequate and realistic provision for the self-weight can be made.

In including the self-weight of the reinforced concrete element in the design calculations, no attempt is made to assess accurately the weight of reinforcement independently of the weight of the concrete; a general density of reinforced concrete is assumed, usually in the order of 2400 kg/m^3. Thus, for a uniform section, the effects of the self-weight are those of a uniformly distributed load over the full span of the beam, and the moments produced may be easily calculated. For a member of varying cross-sectional shape the self-weight effects are slightly more difficult to determine, as the equivalent load intensity becomes non-linear, but the extra complexity of calculation lies in the arithmetic only, the basic procedure being the same as for the uniform section.

2.6   Deflection of beams

The deflection of beams designed by the modular ratio method is not usually a critical factor in the design, although investigation of this serviceability condition is of greater importance in beams designed by the limit state approach, and in certain classes of prestressed section. In all sections, the deflection should be limited in order to avoid damage to services and finishes. Estimation of the exact deflection of reinforced concrete beams is difficult as there are a number of factors that may have considerable effect on the deflection, but which are difficult to assess. These include:

(i) The long-term effects (creep) of the concrete.
(ii) The effects of finishes, partition walls and other non-structural elements of the structure.
(iii) Shrinkage of the concrete under the setting action.
(iv) The precise duration of the live load: e.g. a railway bridge may be very heavily loaded when the train passes over it, but the duration of the load is short, and frequently only a small percentage of the life of such a structure is spent in the heavily loaded state. This contrasts with a footbridge in which the live load is comparatively low when

compared with the dead load (self-weight) of the structure, so that the bridge is effectively never in a state of low load.

Some estimation of the deflection of a beam may be obtained by assuming linear behaviour of the structure and the materials, when the deflection may be calculated from

$$\frac{M}{E_c I_{ce}} = \frac{d^2 y}{dx^2}$$

where E_c is the modulus of elasticity of the concrete and I_{ce} is the second moment of area of the section in equivalent concrete units. This equation may be developed to give a direct expression for the maximum deflection, such as

$$a = k \frac{L^2 M}{E_c I_{ce}} \qquad (2.15)$$

where k is a coefficient that depends on the load-distribution and the end fixity conditions. For example, for a simply supported beam under a concentrated load in the centre, $k = 1/12$; but this ignores the effects of the self-weight, for which the value of k applicable to a simply supported beam under a uniformly distributed load applies ($k = 5/48$). In practice, deflection is not usually considered as a separate factor in modular ratio design, but is allowed for at the design stage by ensuring that the beam has adequate stiffness to resist deflections, achieved by limiting the span : depth ratio.

2.7 Compression members

The modular ratio technique is not suited for the design of compression members owing to the difficulties in obtaining an accurate indication of stresses when using the elastic theory for reinforced concrete sections. This is largely due to uncertainties in determining the exact value of the modular ratio, which has a much greater effect on the theoretical performance of the column than is the case for beams. The design of compression members is therefore best carried out by the ultimate load method described in Chapter 3.

2.8 Effects of creep and shrinkage

Discussion of the properties of concrete in Chapter 1 showed that the exact measurement of the modulus of elasticity of concrete is complicated by the effects of creep, which for a given stress produces with time a change in the elastic modulus. The effect of this is to cause a corresponding change in the modular ratio, which increases as the effective modulus decreases. For flexural members, an increase in the modular ratio causes an increase in the depth to the neutral axis (eq. (2.8)). Hence eqs. (2.1) and (2.2) show that the effect of creep on the stresses in the section is to reduce the concrete stress and to increase the reinforcement stress. A similar situation may be shown to exist in compression members, i.e. that creep effectively reduces the concrete stress while increasing the stress in the reinforcement.

Except for prestressed members (Chapter 5), no direct provision is usually made to include creep effects in design calculations. Creep effects are considered indirectly, however, in the choice of the modular ratio value used in the design; although the modular ratio for fresh concrete may be as low as 6, and is constantly increasing with age to a value of 30 or more, by taking a ratio of 15 some allowance is made for the creep effects while still ensuring that the concrete is not greatly overstressed in the short term.

Unlike creep, shrinkage is not affected by the stress in the concrete, but depends primarily on the environment surrounding the member, although the composition of the concrete, amount of reinforcement, thickness of the member, and time since casting the section all have some influence. Shrinkage is mainly due to loss of moisture from the structure, for which reason fresh concrete is frequently damped down in the early stages of its life.

No attempt is made in the modular ratio method of design to evaluate shrinkage or shrinkage effects, although some guidance is provided in BS 8110, which deals with the load factor method of design. Since shrinkage causes a reduction in the size of the member, restrained shrinkage produces tension in the member. For most sections the reinforcement that is provided for other reasons is sufficient to carry the extra stress introduced by restrained shrinkage, but nominal reinforcement may be provided in sections in which reinforcement would otherwise be absent.

3 Ultimate or Load Factor Method of Design and Analysis

3.1 Introduction

Chapter 2 showed that the modular ratio method of design and analysis is concerned with limiting the stresses that occur in a structure to certain permissible values, and that calculation of the stresses is carried out assuming a linear stress-strain relationship to hold for the materials concerned. The permissible stresses are related to the ultimate or yield stresses by factors of safety which are intended to provide allowance for material imperfections, design approximations or extra loading.

The factors of safety obviously provide a safeguard against collapse of the structure, and as such might be expected to provide a constant ratio between the working loads and the loads that would cause collapse. In practice it is found that this is not so, and the design may have either too high a factor of safety, giving an uneconomic structure, or one that is too low, resulting in a potentially dangerous structure. Although the factors of safety that were used in the relevant British codes of practice prior to the current issue were such that the former situation was encountered rather than the latter, improvements in the techniques of concrete production leading to higher strength concretes, together with advances in methods of design and analysis, meant that over a period of time a gradual reduction in the factors of safety took place; as the material stresses increased and the safety factors reduced, so methods of design and analysis became more divorced from actual behaviour, leading to the situation where the true factors of safety were virtually unknown.

One of the main reasons for this was the assumption of the linear stress-strain relationship for concrete as used in the modular ratio method, and although CP 114: 1957 was largely based on the modular ratio design concept, it also permitted the use of the load factor method. This was intended to overcome the main disadvantage of the modular ratio method (the fact that the actual factor of safety is not accurately known) by calculating the ultimate load that a given section is able to withstand, and equating this to the working load by a load or safety factor.

The calculation of the ultimate load followed the general procedure that is described later in this chapter, but involved the use of a factor applied to the material stresses in recognition of the fact that some materials have a greater variability in performance than others. In order to present the method in a similar fashion to the well-established modular ratio method, the load factor was applied to the material stresses rather than to the load to be carried; thus, instead of multiplying the load by a factor and equating this to the design load, the stresses permitted in the materials were divided by the factor and the ultimate load of the section calculated from these stresses and then equated to the working load. A result of this presentation was that many students of reinforced concrete design were not readily able to appreciate the inherent advantages of the load factor method or to grasp the true reasons for its use. The acceptance of the philosophy of the design procedure was therefore initially slow.

The design method embodied in CP 110: 1972 *The structural use of concrete* was a refinement of the load factor method in that, although the basic design procedures were the same, the philosophy of the method had undergone considerable development. This philosophy, which also forms the basis of the BS 8110 approach, considers that any structure that is unfit for the use for which it was designed has reached a limit state. The limit state may be reached because the structure is in danger of collapse (limit state of collapse), or because excessive deflection has resulted in the structure's being unable to carry out its design function (limit state of deflection). Other limit states may be reached due to vibration, cracking, durability, fire, or various other factors which mean that the structure can no longer fulfil the purpose for which it was designed.

The principal difference between the philosophy of the limit state method and that of the load factor approach lies in the attitude taken in evaluating the load factors that are used. This involves the use of statistics to examine the variations in the values of the various contributions influencing the limit state of the structure. Fully to understand the implications of the approach it is necessary to have some knowledge of statistics, but a working understanding can be obtained from the information presented below.

It is assumed that the distribution of the variations in material strength and structural loads is represented by the normal or Gaussian distribution. This follows the general form shown in Fig. 3.1, and it is a feature of this type of distribution that, although the number of specimens having a very high or a very low strength is small, it is never zero.

Although the curve shown is for stress at failure, a similar curve may be drawn for loading conditions, which also shows a small but finite possibility that an excessively high or low load may occur. It is therefore theoretically possible to have the situation in which two extremes are reached simultaneously, and if this should be the extremes of high load together with low strength, then a limit state of collapse may be reached.

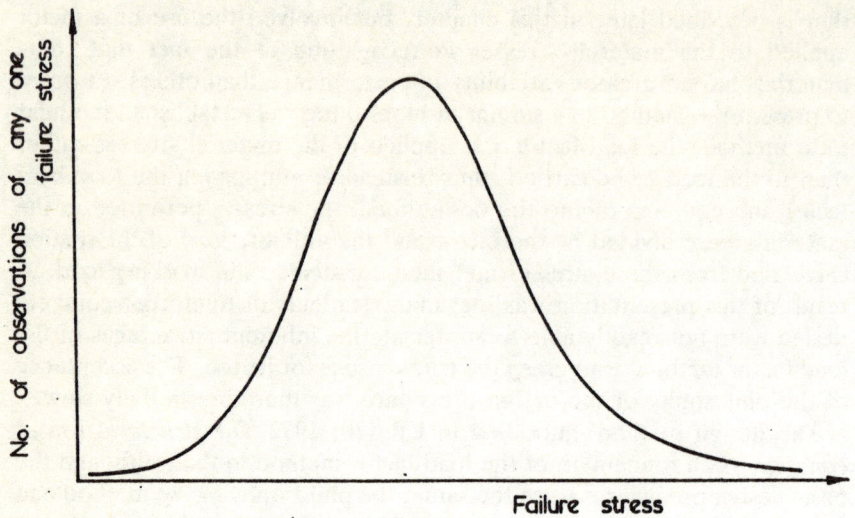

Fig. 3.1 Normal or Gaussian distribution curve.

The probability of such a condition occurring is kept below a chosen figure by selecting suitable design loads and stresses, although the theory does not allow zero probability of the collapse limit state being reached. The effect is that, although no structure can ever be considered absolutely safe, the probability of failure is made sufficiently low that it may be taken as being effectively zero.

The idea of even a small possibility of collapse in a structure being acceptable is one that has led to considerable discussion, and there have been various suggestions as to what should constitute a generally acceptable probability of collapse. In such discussions, however, it is essential that the theoretically possible failure of a structure (together with any resulting death or injury) be viewed in relation to all the risks normal to day-to-day existence in this world, so that some degree of perspective on the subject is reached.

The use of statistical procedures has resulted in the use of characteristic loads and strengths as reference values. The definition of a characteristic load is given (CP 110 handbook) as 'that value of load which has an accepted probability of not being exceeded during the life of the structure', while a characteristic strength (f_k) is defined on the basis of material tests as being

$$f_k = f_m - ks \tag{3.1}$$

where f_m is the mean strength, s is the standard deviation and k is a coefficient that depends on the proportion of test results that will be accepted as having a strength less than the characteristic strength. In BS 8110, k is taken as 1·64. This means that, on a normal distribution, not more than 5 per cent of the test results will fall below the characteristic strength.

Further allowance is made for the effects of variability in the materials by using a partial safety factor γ_m, so that the design strength of a material is given by f_k/γ_m. This factor is introduced to allow for differences that may occur between the strength of the material as determined from a laboratory test and that achieved in the structure. The difference may occur for a number of reasons—method of manufacture, duration of loading (a short-term laboratory test may give a higher strength than is obtained in the structure), corrosion, and other factors.

The values of γ_m for a reinforced concrete structure are generally taken as 1·5 for concrete and 1·15 for the steel reinforcement, except in consideration of excessive short-term loads or localised damage, when the values are reduced to 1·3 and 1·0. The difference in values for the two materials is indicative of the comparative lack of control that may be exercised over the production of concrete, the strength of which is affected by factors such as the water : cement ratio, compaction, rate of drying, etc., which frequently cannot be accurately controlled on the site, when compared with the production of reinforcing steel, which is always carried out under strict control in factory conditions. The values of γ_m given above are for the limit state of collapse: other values of γ_m may be used for alternative limit states.

The same basic procedure may be used for the calculation of characteristic loads in that an ideal relationship may be developed between the characteristic load, the mean load and the standard deviation from the mean, similar to that given as eq. (3.1). Practical considerations, notably insufficient statistical information, reduce the effectiveness of the approach with regard to the prediction of characteristic loads, however, so that the characteristic loads are defined as being:

(i) Characteristic dead load G_k : the weight of the structure complete with finishes, fixtures and partitions, taken as being equal to the dead load as defined in BS 6399.
(ii) Characteristic imposed load Q_k : the imposed load as defined in BS 6399.
(iii) Characteristic wind load W_k : the wind load as defined in CP 3: Chapter V: Part 2.

The design load that arises from consideration of the various characteristic loads is determined by applying a partial safety factor to the characteristic load values. This load factor, γ_f, is introduced to take into account possible increases of load above those considered at the design stage, inaccuracies in the analysis and design procedure, variations in the physical dimensions of the member due to constructional procedures, and the importance of the limit state that is being considered. The design load therefore becomes $\gamma_f F_k$ where F_k is the appropriate characteristic load. The values of γ_f to be used are given in BS 8110 and are summarized in Table 3.1.

For the limit state of collapse, therefore, the design load for a structure under a combination of dead load, imposed load and wind load becomes

$$1 \cdot 2 G_k + 1 \cdot 2 Q_k + 1 \cdot 2 W_k$$

The factor of safety of the structure is given by the product $\gamma_f \gamma_m$, so that for the above condition the factor of safety is $1 \cdot 2 \gamma_m$, i.e. $1 \cdot 38$ for steel and $1 \cdot 8$ for concrete. Although this may appear to be low, the probability of an overload situation being achieved in all three of the load conditions simultaneously is exceedingly remote, so that the effective factor of safety is somewhat higher.

Table 3.1   Values of γ_f for various load conditions

Load conditions	Design dead load	Design imposed load	Design wind load
Dead and imposed load	$1 \cdot 4 \ G_k$	$1 \cdot 6 \ Q_k$	—
Dead and wind load	$1 \cdot 4 \ G_k$	—	$1 \cdot 4 \ W_k$
Dead, imposed and wind load	$1 \cdot 2 \ G_k$	$1 \cdot 2 \ Q_k$	$1 \cdot 2 \ W_k$

G_k, Q_k and W_k are the characteristic dead, imposed and wind loads respectively.

3.2   Basis of the method

In referring to the limit state approach, it must be emphasised that this is a philosophy of design and not a design method in itself. The method of design that is associated with the limit state approach is known as the 'ultimate load method' in which the behaviour of the materials up to the point of failure is considered. The method takes account of any non-linearity in the behaviour of the materials, and thus shows a much closer correlation with the true state of affairs than does the elastic or modular ratio approach. However, there is one important deficiency in using this procedure: since the technique is concerned with the behaviour of the section at ultimate load, and is in effect attempting to predict the ultimate load, it is unable to make any assessment of the behaviour of the structure at working loads. Should such an assessment be required, for example to estimate the deflection of the structure, it is therefore necessary to turn to the techniques based on elastic methods.

The development of ultimate strength theory depends on three assumptions which are not specifically restricted to ultimate strength theory, but also apply to the modular ratio method of design and analysis. These are as follows:

(i) Plane sections remain plane.
(ii) Concrete has zero tensile strength.
(iii) The stress-strain relationship for both concrete and steel in flexure is that determined by standard tensile and compressive tests.

The design procedure follows four basic steps:

1. Obtain the design load by multiplying the characteristic loads by the partial safety factors.
2. Apply these loads to the structure and assume that the structure is at the point of collapse.
3. Calculate the moments and forces at failure.
4. Design the section to carry the stresses produced by Step 3.

We are here concerned with the last of these steps only, the design of the actual section, but Steps 1 to 3 also have their place in the overall design method.

3.3   Sections in flexure

The general procedure for developing the equations of analysis and design is very similar to that described in Chapter 2, i.e. the longitudinal compressive and tensile forces are equated, and the moment of resistance of the section is equated to the applied bending moment. The main difference lies in the distribution of the stresses in the concrete, and this is illustrated in Fig. 3.2 which shows the effect of increasing bending moment on the concrete stresses in a rectangular singly reinforced section. Figure 3.2(b) shows the situation when the moment is quite small and the tensile stress in the section is less than that causing cracking in the concrete.

As the moment is increased, however, the situation shown in Fig. 3.2(c) is reached. At this bending moment the concrete in the tension zone has cracked, so that all of the tensile forces are carried by the reinforcement, which at this stage has not yet yielded. Although the strain is increasing at a linear rate, the increase of the stresses in the concrete follows the parabolic form of the stress-strain relationship for concrete under compression that was shown in Fig. 1.3.

A further increase in the bending moment causes yielding of the reinforcement, and produces the compressive stress distribution shown in Fig. 3.2(d), the behaviour of concrete in compression producing the stress distribution that leads to the maximum stress in the concrete occurring at

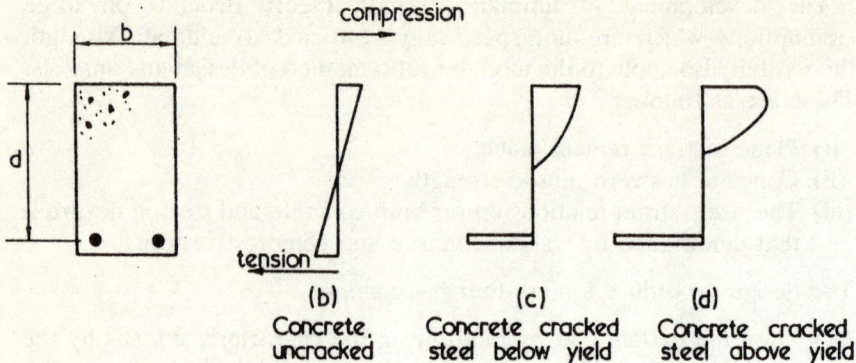

(b)
Concrete,
uncracked

(c)
Concrete cracked
steel below yield

(d)
Concrete cracked
steel above yield

Fig. 3.2 Stress-distribution in an under-reinforced section under flexure.

a level other than the extreme fibre. Still further increase of the bending moment produces failure of the section, either by fracture of the reinforcement or by crushing of the concrete.

Examination of Fig. 3.2 shows that the equilibrium of the longitudinal forces in the section is maintained by virtue of a change in the position of the neutral axis. If the section is under-reinforced, as shown in Fig. 3.2, the neutral axis moves towards the compression face, but if it is over-reinforced the neutral axis moves towards the tension face. The reason for this is that, in the former case, the reinforcement yields before the section fails. Increasing the bending moment is not therefore accompanied by an increase in the tensile forces in the section, although it causes an increase in the compressive stress level. In order to maintain the equilibrium between the compressive force in the section and the tensile force in the section, therefore, the compressive stress must act on a reduced area. This is achieved by the movement of the neutral axis towards the compressive face, and has the effect of maintaining the force but increasing the lever arm, so that the additional bending moment may be carried. If the bending moment is further increased, the beam shows considerable deflection, producing high concrete strains in the extreme fibre, which are accompanied by cracking in the concrete and spalling of the concrete from the compressive face. This reduces the depth of the section, resulting in the failure of the beam. The load-deflection behaviour of an under-reinforced beam is shown in Fig. 3.3.

If the section is over-reinforced, the concrete crushes before the reinforcement reaches yield stress. This is both uneconomic, since the material is not being used to its full advantage, and also potentially dangerous, as the section never exhibits the high deflection that characterises the third phase of the behaviour shown in Fig. 3.2(d); collapse occurs without warning.

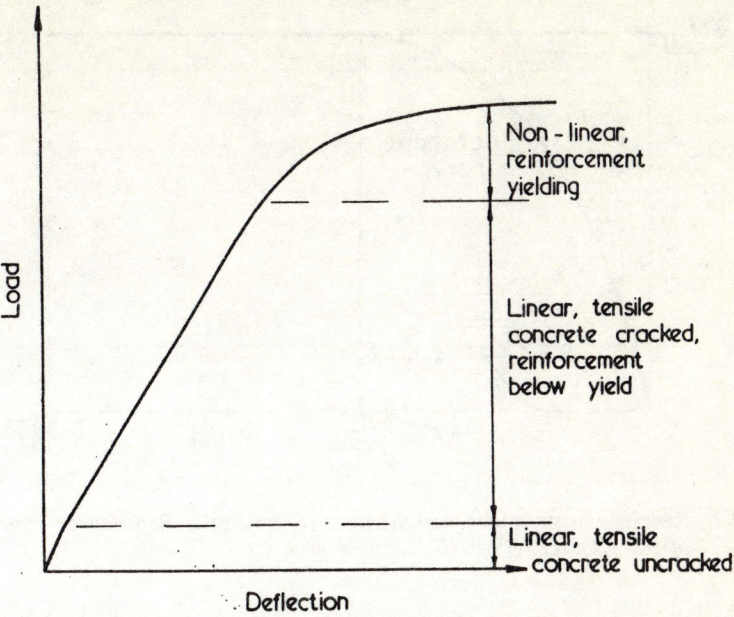

Fig. 3.3   Load-deflection behaviour of an under-reinforced section in flexure.

The actual stress-strain curve for concrete under compression was shown in Fig. 1.3. It is extremely difficult, however, to take account of the falling-branch part of the curve in developing design equations or in analysing the performance of a section, so that for design purposes the stress-strain relationship for concrete in compression may be taken as being a rectangular parabola with a straight line extension, as shown in Fig. 3.4.

Development of the design equations requires knowledge of the total compressive force together with the line of action of that force. For rectangular sections these two parameters are easily obtained from the rectangular parabolic curve, but if the section is more complex in cross-sectional shape, application of the rectangular paraboloid stress-strain curve becomes somewhat tedious and it is much more convenient to consider alternative stress blocks. Any stress-distribution may be used provided that the total compressive force as calculated from the distribution, together with the line of action of that force, is compatible with the 'true' distribution.

Many suggestions have been made for equivalent stress-distributions, the most convenient being a rectangular stress block in which the stresses are assumed to be constant over a certain depth. BS 8110 suggests the use

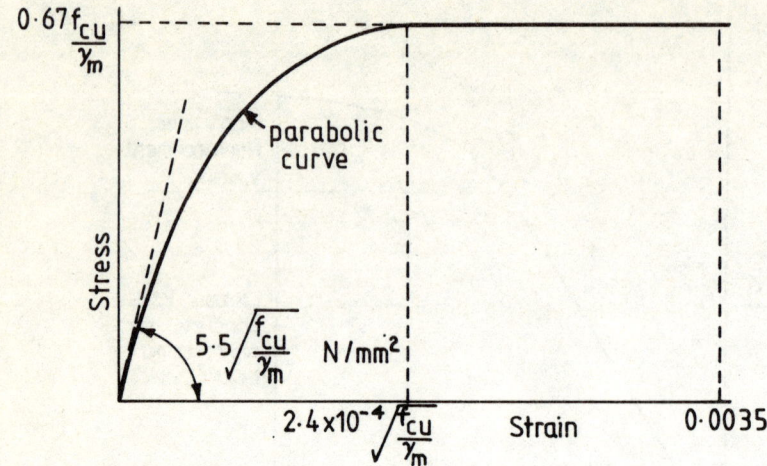

Fig. 3.4 Rectangular-parabolic design curve for concrete. *Reproduced courtesy of BSI. Source: BS 8110 Part 1, Figure 2.1.*

of the rectangular stress block extending over 0·9 of the full depth of the compression zone. The value of the maximum stress at flexural failure is taken as being $0·67f_{cu}$; after the partial safety factor γ_m has been applied, the design stress therefore becomes $0.45f_{cu}$ over the complete compression zone.

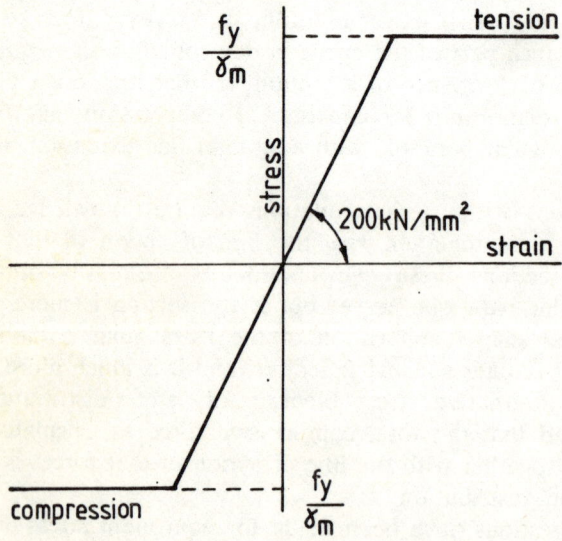

Fig. 3.5 Design stress-strain curve for reinforcement. *Reproduced courtesy of BSI. Source: BS 8110 Part 1, Figure 2.2.*

Although this assumption looks to be rather sweeping, the use of the reduced depth rectangular stress block produces a final result that is very close to that obtained by using the rectangular parabolic stress distribution, and is, of course, much easier to use. A comparison between the values of total compressive force and lever arm for the two stress distributions is given in the Appendix.

The distribution of the tensile stresses does not produce the same problem since, as the stress is concentrated over a small area (the reinforcement area), it may be assumed to be uniform over the whole area. The stress-strain relationship shown in Fig. 1.6 is, however, modified to give a more simple design curve; Fig. 3.5 shows the stress-strain curve for hot rolled reinforcement that is assumed for design purposes. It will be seen that although the design curve clearly shows the change from elastic to plastic behaviour, no allowance is made for work hardening after yield, and that the stress-strain relationship is assumed to be the same for steel in compression as it is for steel in tension.

3.3.1 Moment redistribution and restrictions in the depth of the compression block

The adoption of the limit state philosophy has implications that extend beyond the area of design only, if the structure is being designed to the limit state of collapse, it is reasonable to expect that it should also be analysed to the same limit state. A detailed treatment of the concepts involved is beyond the scope of this text, but the basic procedure is as follows.

When a moment is applied to a beam, the beam deflects, the amount of deflection being linked to the moment by the $d^2y/dx^2 = M/EI$ equation that is derived in books on structural analysis or structural mechanics. This equation, which describes a linear relationship between moment and deflection, holds as long as the beam is under elastic conditions, but if the material has a non-linear stress strain relationship, or is stressed beyond the point at which linear behaviour results, then the moment/deflection relationship also departs from the linear. The exact moment/deflection curve for a reinforced concrete section is very difficult to produce. Not only does the concrete not have a linear stress-strain behaviour, but the cracking of the concrete in the tensile stress zone effectively changes the apparent second moment of area of the section, hence leading to a complex moment/deflection relationship. It is therefore convenient to assume that the graph of moment/deflection is as shown in Fig. 3.6, and involves an initial linear portion followed by a plateau, indicating that once moment has reached a certain value then the beam will deflect greatly with no increase in moment. This occurs at the collapse moment M_u, and implies that for a homogeneous section the material of the beam is at yield throughout and that a hinge has developed. Even though the material of a reinforced concrete section will not yield in the accepted sense, a hinge will still effectively form.

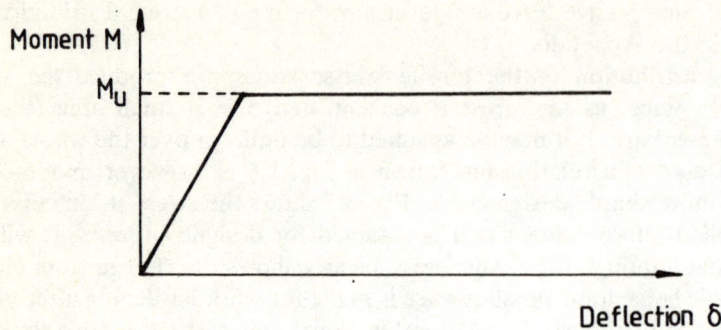

Fig. 3.6 Assumed moment/deflection relationship.

Although Fig. 3.6 is in terms of moment/deflection, deflection can be considered analogous to rotation or curvature, and the relationship between moment and curvature or moment and rotation can be taken to be the same as that shown in Fig. 3.6 for moment and deflection. The implication of this is that there is an ultimate moment which cannot be exceeded, and that once this moment is reached, a hinge will form producing unlimited rotation. Of course, the rotation has a limit in practice, since when the strain levels reach certain physical limits failure of the material will occur, and buckling may also affect the behaviour, but ignoring these factors enables the effect to be illustrated with a simple example.

Consider a fixed end beam having a uniformly distributed load of $w/$ unit length. The elastic bending moment is as shown in Fig. 3.7, with support moments that are twice the mid-span moment.

As the load is increased, so do the moments and rotations, until the load reaches w_1 at which point the moments at the supports A and C are $w_1 l^2/12$. Assume that these moments are equal to the collapse or ultimate moment M_u. Any increase in load cannot cause an increase in moment above M_u (Fig. 3.6), and in effect hinges, allowing (for all practical purposes) unlimited rotation, have formed at A and C. The beam will not collapse as it is not a mechanism, but any further increase in load will cause rotation at these hinges with no further increase of support moment — i.e. for loads above w_1 the beam will behave as if it were simply supported. Hence a load of intensity $(w_1 + w_2)$, where w_2 is the increase in load over w_1, will produce a BMD as shown in Fig. 3.8, and if this load is increased, so the moment at point B will increase until the total moment at B arising from loads w_1 and w_2 reaches M_u.

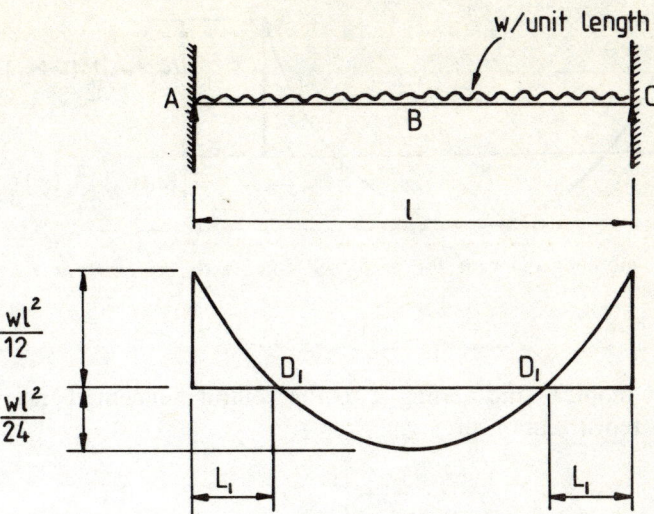

Fig. 3.7 Elastic bending moment diagram for fixed end beam under a uniformly distributed load.

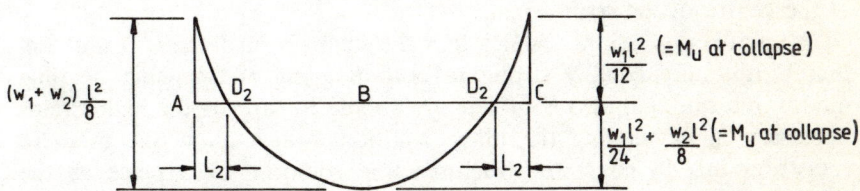

Fig. 3.8 Bending moment distribution at plastic collapse.

At this stage a hinge will form at B, and the beam will collapse, since it has now become a mechanism. At collapse the BMD is as shown in Fig. 3.8, with M_A and $M_C = w_1 l^2/12$, and $M_B = (w_1 l^2/24 + w_2 l^2/8)$ and it is interesting to compare this moment diagram with that which would be obtained if the structure has been assumed to behave in an elastic fashion up to failure with no premature hinge formation at A and C (Fig. 3.9).

Notice that the total range of moment between A and B is the same for both causes, but in Fig. 3.9 the support moment is greater than M_u, which according to Fig. 3.6 cannot occur. To arrive at the moment diagram shown in Fig. 3.8, some redistribution of the elastic bending moments as shown in Fig. 3.9 are required. This is achieved by subtracting $w_1 l^2/12$

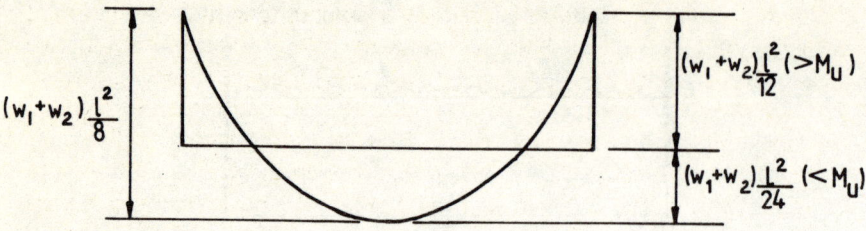

Fig. 3.9 Bending moment diagram for collapsed load with no moment re-distribution.

from the support moment and adding it to the central moment, hence there has been a redistribution of

$$\frac{w_2 l^2/12}{(w_1 + w_2)l^2/12}$$

at the supports, and

$$\frac{w_2 l^2/12 + w_2 l^2/8}{(w_1 + w_2)l^2/24}$$

at the centre of the span.

The above discussion assumes that the beam is made from a material that is not only equally strong in both hogging and sagging bending modes, but one that also is sufficiently ductile to support the high strains necessary in developing the hinge rotations which allow the moment redistribution. In concrete structures this rotation takes place as the reinforcement yields, but if the section is over-reinforced, the amount of plastic deformation (rotation) is greatly limited since the concrete is unable to accommodate the strain required to develop the hinge. In order to be certain of a satisfactory plastic hinge development, the section must therefore be under-reinforced. Consideration of the strain distribution across the section shows that this is ensured by limiting the depth of the compression zone to a maximum of $0.5d$, for which value a moment redistribution of 10 per cent or less is assumed. If the moment redistribution exceeds 10 per cent, the neutral axis depth must be reduced in order that the rotational capacity of the section is increased to allow for the extra redistribution, and BS 8110 introduces the condition that $x \not> (\beta_b - 0.4)d$, where β_b is given from

$$\beta_b = \frac{\text{moment at section after redistribution}}{\text{moment at section before redistribution}}$$

As outlined below, although not specified directly, the amount of moment redistribution must not exceed 30 per cent, so that in effect $\beta_b > 0.7$.

The moment redistribution procedure is concerned with the situation at collapse, and does not concern the moment states at working or service loads, which would be represented by a BMD such as that shown in Fig. 3.7, with w as the service load. Under these conditions, the point of contraflexure occurs at D_1 – distance L_1 from the support points. However, taking the moment to collapse, and redistributing, produces the moment diagram shown in Fig. 3.8, where it can be seen, the point of contraflexure has been moved to D_2 – distance L_2 from the supports ($L_2 < L_1$). Fig. 3.8 is the ultimate design condition, and it may be appreciated that designing for this condition implies designing for no moment at point D_2. However, under the service conditions as represented by Fig. 3.7, moments will occur at this point, and if suitable reinforcement is not provided, then cracks, and even premature failure may occur. In order to allow for this situation, BS 8110 includes the requirement that the moment of resistance of the beam at any point must not be less than 70 per cent of the maximum elastic moment that occurs at that section under the ultimate load conditions, but before redistribution. In effect, this means that the amount of moment redistribution must not exceed 30 per cent, and leads to a design BMD that may show unexpected discontinuities. The example below illustrates this point.

Although the moment redistribution procedure outlined above has been presented for a single span beam, it should not be thought that the procedure is restricted to single span conditions. The process may be equally well applied to multi-span structures, but the overriding essential feature of all moment redistribution is that the elastic moment diagram is modified. In multi-span beams a complication arises in that the application of the load to the various spans may produce a large number of load cases, thus complicating the construction of the final redistributed moment diagram which is used in the design procedure.

Example
Draw the design bending moment diagram for the beam shown in Fig. 3.10(a), assuming that a moment redistribution of 25 per cent occurs.

The elastic bending moment diagram and the point of contraflexure may easily be found as the beam is statically determinate (Fig. 3.10(b)). Applying a moment redistribution of 25 per cent produces the BMD shown in Fig. 3.10(c), the important element of this diagram being that the total range of moment in both Figs. 3.10(b) and 3.10(c) are the same, showing that the system is in moment equilibrium. The point of contraflexure for this redistributed condition has moved however, and is now 1·375 m from the supports. At working or service loads, the bending moment at this point is 52·5 kNm, and since the moment of resistance should be at least 70 per cent of the elastic moment, this means that a

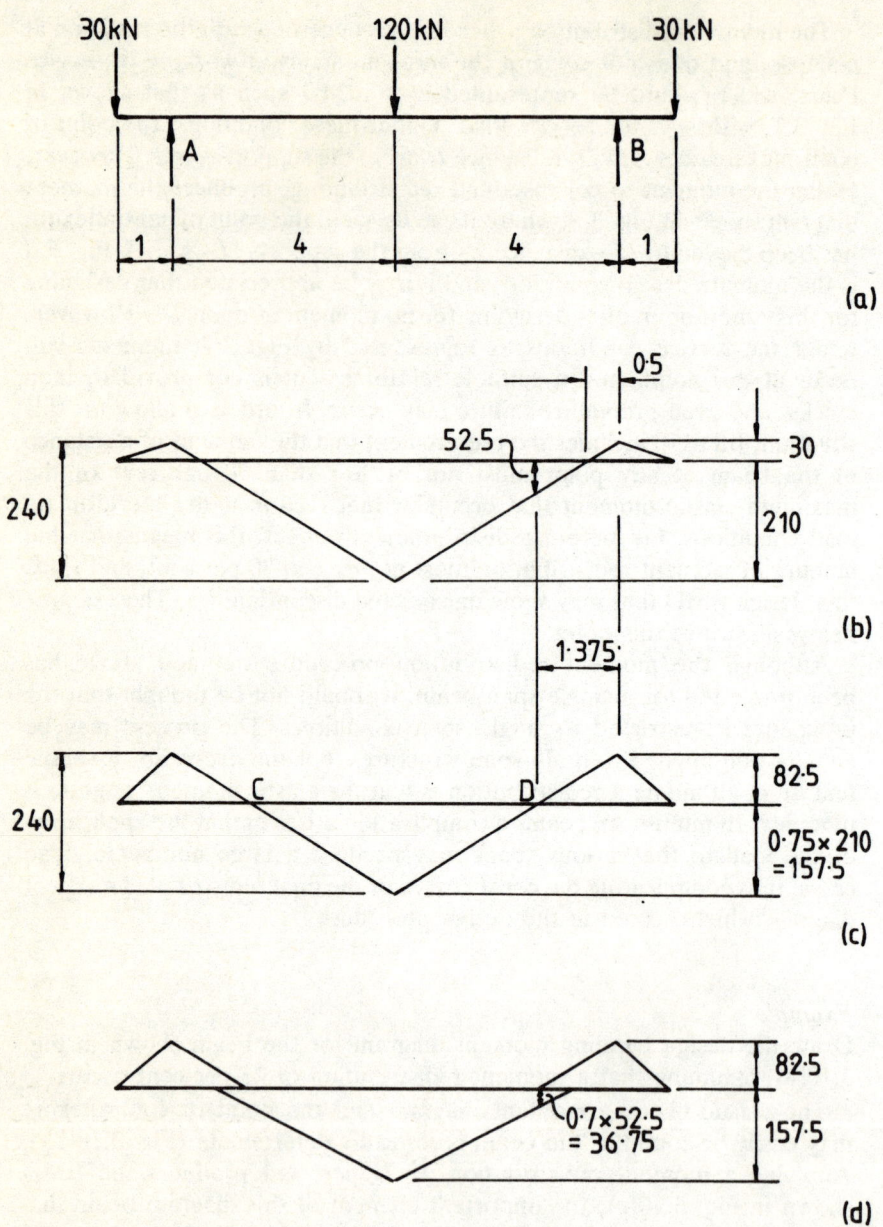

All bending moments are in kN m

Fig. 3.10

moment of resistance of at least $0.7 \times 52.5 = 36.75$ kNm should be provided at points C and D, thus producing the design moment diagram shown in Fig. 3.10(d).

The equations developed below apply to the situation where the re-distribution of moment is 10 per cent or less. Equations relevant to greater moment-distribution are developed in exactly the same way, the only difference in the equations stemming from the reduction in the depth of the compression zone.

3.4 Singly reinforced sections

Consider the singly reinforced section shown in Fig. 3.11. The force in the compression zone

$$F_c = 0.45 f_{cu} b \times 0.9x$$
$$= 0.4 f_{cu} bx \tag{3.2}$$

and acts through a point $0.45x$ down from the extreme compression face. Equating the compressive and tensile forces,

$$\frac{A_s f_y}{1.15} = 0.4 f_{cu} bx$$

therefore

$$x = \frac{A_s f_y}{0.46 f_{cu} b} \tag{3.3}$$

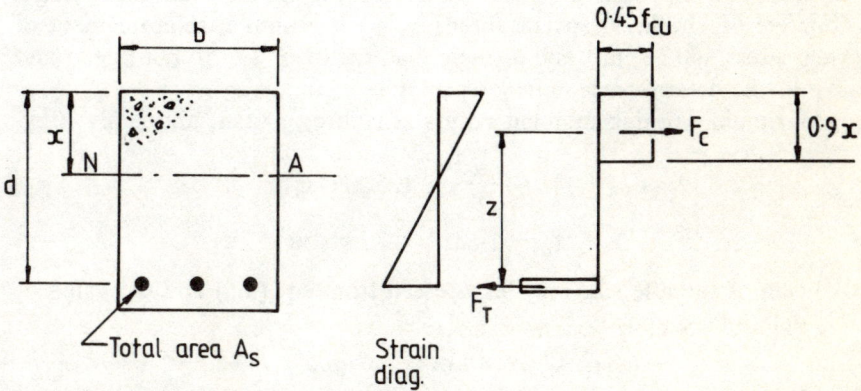

Fig. 3.11   Singly reinforced section under flexure.

The lever arm

$$z = d - 0\cdot45x = d - 0\cdot98\frac{A_s f_y}{f_{cu} b} \qquad (3.4)$$

so that, taking moments about the line of action of the tensile forces,

$$M = 0\cdot4 f_{cu} bx \left(d - \frac{A_s f_y}{f_{cu} b}\right) \qquad (3.5)$$

The maximum moment that can be carried by the singly reinforced section is governed by the necessity of ensuring that the reinforcement yields before the concrete crushes, and is obtained by limiting the depth to the neutral axis to $d/2$. This means that there is an absolute value of moment, above which the singly reinforced section cannot be taken. In the limit, therefore, $z = 0\cdot775d$ when $x = d/2$, so that the ultimate moment is given, in terms of concrete stress, by

$$M_u = 0\cdot4 f_{cu} b\,\frac{d}{2} \times 0\cdot775d = 0\cdot155 f_{cu} bd^2 \qquad (3.6)$$

or, working in terms of steel stress,

$$M_u = \frac{A_s f_y}{1\cdot15} 0\cdot775d = 0\cdot67 A_s f_y d \qquad (3.7)$$

Equations (3.6) and (3.7) give the absolute maximum moment of resistance of the singly reinforced section, and provide enough information for the section to be sized and the amount of reinforcement to be determined.

Example 3.1
A singly reinforced concrete beam spans 6 m and carries uniformly distributed dead and live loads of 30 kN/m over the full span. Assuming a concrete of Grade 30 specification ($f_{cu} = 30$ N/mm^2), reinforcement of yield stress 460 N/mm^2 and a single load factor of 1·6 on both dead and live loads, determine a suitable section.

Maximum working moment occurs at centre of span, and is given by

$$M = \frac{\omega L^2}{8} = 135 \text{ kN m}$$

$$M_u = 1\cdot6M = 216 \text{ kN m}$$

A beam of suitable size may be assessed from eq. (3.6) and the value of M_u determined above:

$$216 \times 10^6 = 0\cdot155 \times 30 \times bd^2$$

therefore

$$bd^2 = 46\cdot5 \times 10^6 \text{ mm}^3$$

Suppose

$$b = \frac{d}{2}$$

then

$$\frac{d^3}{2} = 46.5 \times 10^6 \text{ mm}^3$$

therefore

$$d = 453 \text{ mm} \quad \text{and} \quad b = 227 \text{ mm}$$

This gives the effective depth; allowing for thickness of reinforcement and cover to reinforcement, the actual depth should be 500 mm and the breadth 230 mm. The reinforcement area required may be obtained from eq. (3.7):

$$A_s = \frac{M_u}{0.67 f_y d} = \frac{216 \times 10^6}{0.67 \times 460 \times 453} = 1547 \text{ mm}^2$$

3.4.1 Design graphs for singly reinforced sections

Example 3.1 is an extremely simple illustration of the basic form of calculation that is required. Although the calculation involves little arithmetical computation, there are occasions, particularly in the analysis or comparison of existing sections, when suitable design graphs prove of considerable benefit. Although such graphs may be developed from the equations already derived, they are of such widespread use that a complete understanding of their derivation is desirable, and development of the graphs is therefore followed from first principles.

Let

$$x = k_1 d \quad \text{and} \quad z = k_2 d$$

Assuming that a rectangular stress block is used for concrete in compression (Fig. 3.11).

$$z = d - 0.45x$$

therefore

$$k_1 = 2.2(1 - k_2) \tag{3.8}$$

Taking moments about the line of action of the tensile forces (as for development of eq. (3.5)) gives

$$M_u = 0.45 f_{cu} b 0.9 k_1 d k_2 d = 0.4 f_{cu} b d^2 2.2(1 - k_2) k_2$$

therefore

$$\frac{M_u}{bd^2} = 0.9 f_{cu} k_2 (1 - k_2) \tag{3.9}$$

Equating the compressive and tensile forces,

$$\frac{A_s f_y}{1.15} = 0.45 f_{cu} b 0.9 k_1 d$$

Rearranging and substituting for k_1,

$$\frac{A_s f_y}{bd} = 1.03 f_{cu} (1 - k_2) \tag{3.10}$$

Substituting into eqs. (3.9) and (3.10) for f_y and f_{cu}, remembering that k_2 must have values between 0.775 and 1.0, allows values of A_s/bd and M_u/bd^2 to be determined (Table 3.2). From these values may be drawn the graph of M_u/bd^2 against A_s/bd for the relevant grades of concrete and steel. The graph for the results given in Table 3.2 is shown in Fig. 3.12.

Table 3.2   Values of M_u/bd^2 and A_s/bd for various values of k_2

k_2	0.775	0.8	0.85	0.9	0.95	1.0
M_u/bd^2	3.92	3.6	2.87	2.02	1.07	0.0
A_s/bd	0.0234	0.0208	0.0156	0.0104	0.0052	0.0

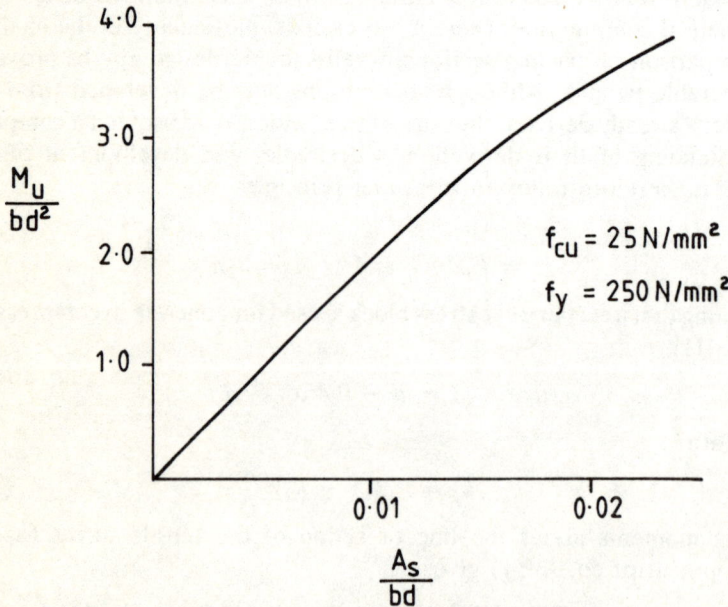

Fig. 3.12   Design graph for singly reinforced beams.

A family of such curves for various values of f_{cu} and f_y is presented in Part 3 of BS 8110. The graphs differ slightly from those obtained by the above procedure, however, as the code uses the rectangular parabolic stress-distribution instead of the linear rectangular distribution used here.

Use of the graphs is very simple, and is illustrated by the example below.

Example 3.2

Determine the ultimate moment that can be carried by the section obtained in Example 3.1, if the concrete is to Grade 25 specification and the reinforcement is mild steel having $f_y = 250$ N/mm^2.

From Example 3.1, $A_s = 1547$ mm^2, $b = 230$ mm, d (after allowing for cover and reinforcement size) $= 450$ mm. Therefore

$$\frac{A_s}{bd} = \frac{1547}{450 \times 230} = 0.0149$$

From Fig. 3.12, the value of M_u/bd^2 corresponding to the above value of A_s/bd is 2.79 N/mm^2. Therefore

$$M_u = 2.79 \times 450^2 \times 230 = 130 \text{ kN m}$$

Therefore, if the material specifications of the section are changed as specified above, the ultimate moment drops to 130 kN m.

The main disadvantage of the graphs obtained by the above process is that each curve is only applicable to one grade of concrete and one grade of reinforcement. The graphs presented in BS 8110 place curves for several grades of concrete on each graph, but still require a number of graphs to cover the combinations of reinforcement and material specification. A simpler alternative, requiring only one graph for all of the various material specifications, is that developed below.

From eq. (3.9),

$$\frac{M_u}{f_{cu}bd^2} = 0.9k_2(1 - k_2) \tag{3.11}$$

Hence the graph of k_2 against $M_u/(f_{cu}bd^2)$ may be constructed (Fig. 3.13). Note that, as before, $0.775 < k_2 < 1.0$, and that the maximum value of $M_u/(f_{cu}bd^2)$ is 0.155 (which agrees with eq. (3.6)). The procedure for using this graph in design is therefore:

1. Choose a section such that $M_u/(f_{cu}bd^2) \not> 0.155$
2. From Fig. 3.13 read off k_2
3. Calculate A_s from the equation

$$A_s = \frac{1.15M_u}{f_y k_2 d} \tag{3.12}$$

which is eq. (3.7), but with the lever arm of $0.775d$ replaced by k_2d. The use of this graph is demonstrated in Example 3.3.

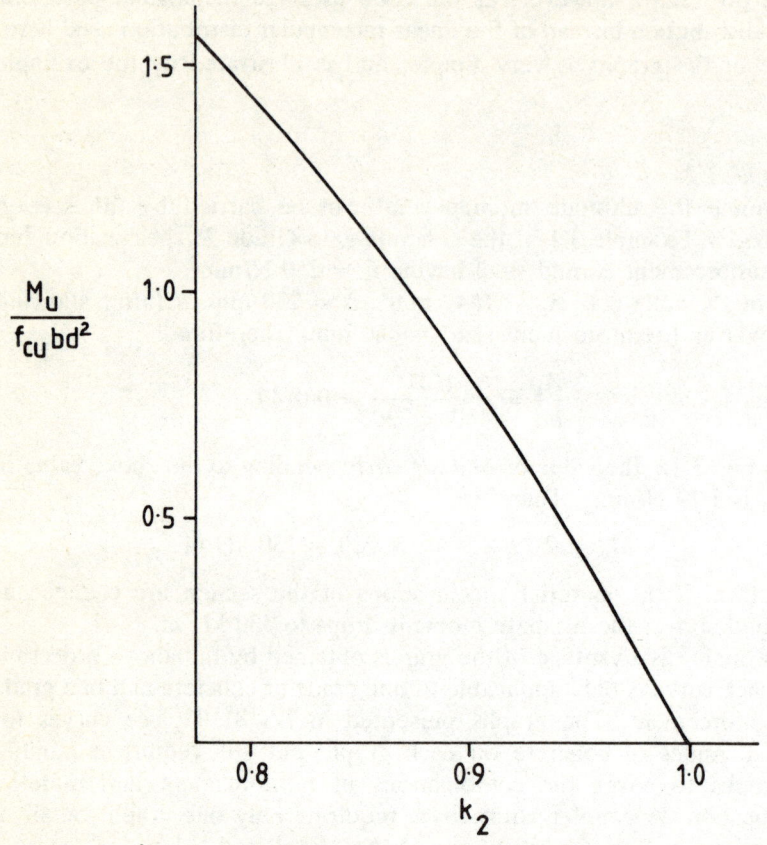

Fig. 3.13 Alternative design graph for singly reinforced beams.

Example 3.3
Repeat the design of a section to carry the loads given in Example 3.1, but using the design graph shown in Fig. 3.13.

As in the previous example, $M_u = 216$ kN m. Taking b 250 mm and $d = 450$ mm,

$$\frac{M_u}{f_{cu}bd^2} = \frac{216 \times 10^6}{30 \times 250 \times 450^2} = 0.142$$

From Fig. 3.8,

$$\frac{M_u}{f_{cu}bd^2} = 0.142 \qquad k_2 = 0.8$$

Therefore, from eq. (3.12),

$$A_s = \frac{216 \times 10^6 \times 1\cdot15}{460 \times 0\cdot8 \times 450} = 1500 \text{ mm}^2$$

This calculation shows the simplicity with which this graph may be used, and it is the author's opinion that as a design aid this graph is preferable to those presented in BS 8110, which are of the general form shown in Fig. 3.12. The BS 8110 curves are, however, superior for analysis of existing sections, so that the recommendation might be to use Fig. 3.13 for design and to use Fig. 3.12 or the corresponding BS 8110 curves for analysis.

3.5 Doubly reinforced sections

The absolute maximum moment that a singly reinforced section can support was shown, in the development of eqs. (3.6) and (3.7), to be limited by the need to ensure that an under-reinforced type of behaviour was obtained. Any substantial increase in the moment capacity of the section can therefore only be realised by the addition of reinforcement in the compression zone, together with the placing of additional tensile reinforcement to balance the longitudinal forces. This has the effect of increasing the moment capacity of the section by means of the couple formed by the forces in the compression reinforcement and the balancing tensile reinforcement, in just the same way as was considered for the modular ratio case in Chapter 2. If the section is initially heavily over-reinforced, no additional tensile reinforcement may be needed.

The usual design procedure is to design the balanced section with a moment of resistance capacity of M_1, and then to include extra reinforcement in both tension and compression to carry the remaining moment M_2. The ultimate value of M_1 is given by eq. (3.6):

$$M_1 = 0\cdot155bd^2f_{cu} \qquad \text{(3.6 bis)}$$

The remaining moment M_2 may be evaluated by taking moments about the line of action of the tensile reinforcement:

$$M_2 = A_s'f_{sc}(d - d')$$
$$= \frac{A_s'f_y}{1\cdot15}(d - d') \qquad \text{(3.13)}$$

which allows the compression steel to be determined. The amount of tensile reinforcement required may be found by equating longitudinal forces in the section:

$$0{\cdot}4f_{cu}b\frac{d}{2} + \frac{A_s'f_y}{1{\cdot}15} = \frac{A_sf_y}{1{\cdot}15}$$

therefore

$$A_s = 0{\cdot}23\,\frac{f_{cu}}{f_y}\,bd + A_s' \tag{3.14}$$

The total moment on the section $(M_1 + M_2)$ is therefore given by adding eqs. (3.6) and (3.13):

$$M_u = M_1 + M_2 = 0{\cdot}155bd^2f_{cu} + 0{\cdot}87A_s'f_y(d - d') \tag{3.15}$$

so that the design procedure may be summarised as:

1. Determine the 'balanced' moment M_1.
2. Determine the excess moment $M_2 = M_u - M_1$.
3. Calculate the compression reinforcement and the total tension reinforcement required (eqs. (3.13) and (3.14)).

Inspection of the above analysis shows that no allowance has been made for the effects of loss in moment due to the replacement of some of the concrete by the compression reinforcement. The effect of this is so small when compared to the overall compressive forces that it may safely be ignored in ultimate load design; in the modular ratio procedure the effect is rather more prominent and is taken into account.

The validity of eqs. (3.13) and (3.14) depends on the development of the full compressive force in the compression reinforcement. This implies that the reinforcement must be stressed to yield stress, which can only be achieved if the strain in the reinforcement is equal to or greater than yield strain. The implication of this is that the ratio d'/d must not be greater than approximately 0·2, for if this value is exceeded the limiting strain is reached in the concrete extreme fibre before the compression reinforcement reaches yield strain. If this condition exists, development of the equations becomes more complex since it becomes necessary to refer to the complete stress-strain curve for the reinforcement in order to determine the reinforcement stresses, which are below yield.

This discussion of the doubly reinforced section assumes that compression reinforcement is added to the section in order to increase the moment capacity of a beam which would, in its singly reinforced form, be too weak for the moments and loads to be carried. In many instances, however, reinforcement may be included in the compression zone for reasons other than absolute strength. A typical example is the built-in beam under a load system such that the end moments are hogging, while the mid span moment is sagging. This of course means that the tension face is the upper face at the supports, and the lower face at the centre of the span; depending upon the span of the section and the distribution of the loads, the most economical reinforcement detail may be to carry the reinforcement at the supports completely across the span of the section, so that the beam effectively becomes doubly reinforced at the centre of the span. In such cases the equations derived earlier may not apply, since the compressive strength of the concrete may not be fully developed and hence the M_1 condition not reached.

This state of affairs is effectively taken into account by assuming that the depth of the concrete compression zone is limited. From Fig. 3.14,

$$M_u = 0 \cdot 45 f_{cu} b 0 \cdot 9x (d - 0 \cdot 45x) + F_{sc}(d - d') \qquad (3.16)$$

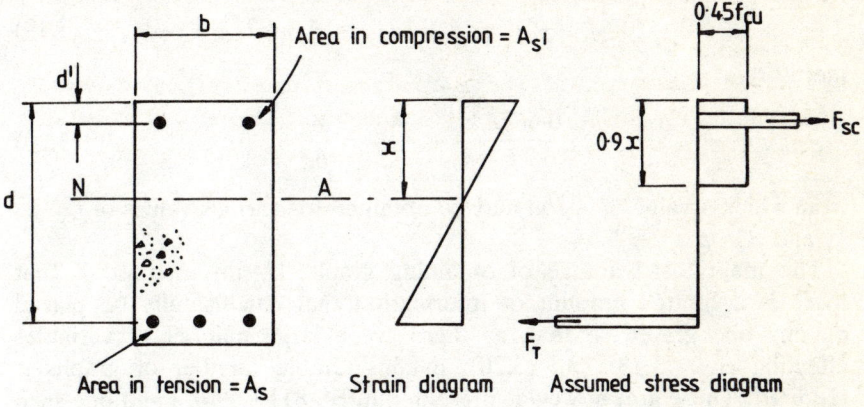

Fig. 3.14   Doubly reinforced section under flexure.

so that for a required value of M_u the position of the neutral axis x may be found. This may then be substituted into eq. (3.14) to give the area of tensile reinforcement required. As before, this approach is only valid if the strain profile of the section is such that the compression reinforcement is at yield. The situation in which this is not so is considered in Section 3.7.

3.5.1   Design graphs for doubly reinforced sections

As with the singly reinforced beam, graphs may be drawn that are a considerable help in the design procedure, but although the method of obtaining the graphs is much the same as that for the singly reinforced beam, the calculations involved are made more complex by the presence of the compression reinforcement. Taking, as before,

$$x = k_1 d \quad \text{and} \quad z = k_2 d$$
$$k_1 = 2 \cdot 22(1 - k_2) \qquad (3.8 \text{ bis})$$

Taking moments about the line of action of the tension reinforcement, we obtain

$$M_u = 0.45f_{cu}b0.9k_1dk_2d + 0.87A'_s f_y(d - d') \qquad (3.17)$$

which is the more general form of eq. (3.15).

From eq. (3.17),

$$\frac{M_u}{bd^2} = 0.9f_{cu}k_2(1 - k_2) + 0.87\frac{A'_s}{bd}f_y\left(1 - \frac{d'}{d}\right) \qquad (3.18)$$

so that, for given values of f_{cu}, f_y. A'_s/bd, d'/d and k_2, M_u/bd^2 may be evaluated. Equating internal longitudinal forces,

$$0.4f_{cu}bk_1d + A'_s 0.87f_y = A_s 0.87f_y \qquad (3.19)$$

therefore

$$\frac{0.46f_{cu}k_1}{f_y} + \frac{A'_s}{bd} = \frac{A_s}{bd} \qquad (3.20)$$

from which a value of A_s/bd may be obtained for various values of f_{cu}, f_y, k_1 and A'_s/bd.

The main disadvantage of obtaining graphs by this method is that there is a limited amount of information that can usefully be placed on any one graph, so that, as there are a large number of variables affecting eqs. (3.18) and (3.20), a considerable number of gràphs is required. These are, however, presented in BS 8110: Part 3 and one such graph is reproduced here as Fig. 3.15.

As with the design charts for singly reinforced sections, the BS 8110 graphs have been developed using the rectangular parabolic stress block. Design charts drawn using equations 3.18 and 3.20 will not therefore produce precisely the same curves as those given in the British Standard.

The use of these graphs is potentially simple, and provides a very quick result. The main difficulty in using them is likely to come from the restriction to the depth of the compression block, which is necessary to allow sufficient rotation of the section for moment redistribution. This is accommodated on the graphs by 'cut-off lines' which depart from the solid graph curve at various points, depending on the x/d ratio. In using the graph, therefore, attention must be paid to the x/d ratio that is applicable to the section in question, and results taken only for ratios that are less than or equal to this. In effect, this means that the relevant portion of the graph is that to the left of the x/d ratio cut-off line, since the cut-off lines are for decreasing ratios from right to left. Example 3.4 illustrates the use of the design graph for a doubly reinforced section.

Example 3.4

A reinforced concrete beam has dimensions restricted to $d = 450$ mm

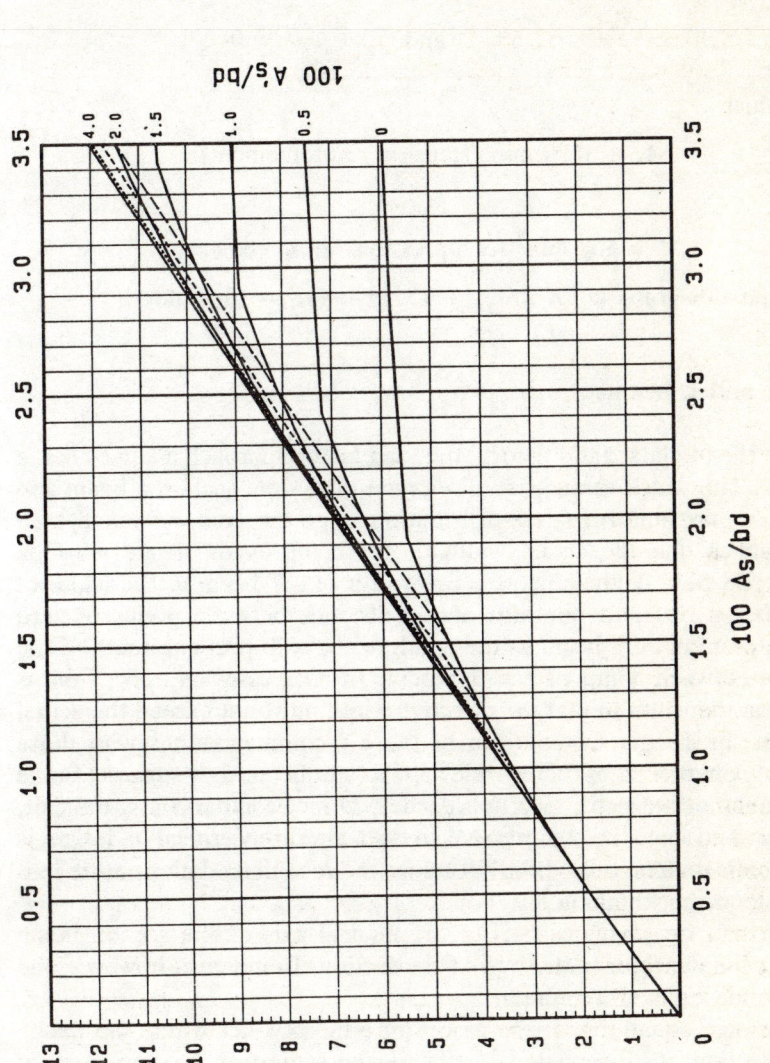

Fig. 3.15 Design graph for doubly reinforced section. *Reproduced courtesy of BSI. Source: BS 8110 Part 3, Chart No. 7.*

and $b = 250$ mm. The ultimate moment to be carried is 280 kN m. Choose suitable reinforcement areas if the concrete is to be to Grade 30 specification ($f_{cu} = 30$ N/mm^2), and the reinforcement has a yield strength of 460 N/mm^2.

$M_u/bd^2 = 5\cdot53$. Assuming that a moment redistribution of 10 per cent is allowed, $x/d \ngtr 0\cdot5$.

From Fig. 3.15, for $M_u/bd^2 = 5\cdot53$ and for $x/d \ngtr 0\cdot5$, one combination of $100A_s/bd$ and $100\ A_s'/bd$ is that giving

$$\frac{100A_s}{bd} = 1\cdot68 \qquad \text{and} \qquad \frac{100A_s'}{bd} = 0\cdot5$$

from which

$$A_s = 1890 \text{ mm}^2 \text{ (tension reinforcement)}$$

and

$$A_s' = 562 \text{ mm}^2 \text{ (compression reinforcement)}$$

Note that this graph is for $d'/d = 0\cdot15$, giving $d' = 67\cdot5$ mm.

3.6   T and L beams

As with the modular ratio theory, the load factor approach assumes that a flanged beam which forms part of a structural system such as a beam and slab floor has uniform stress-distribution over the compression flange. This implies that the actual width of the compression flange must be limited. BS 8110 defines the effective width of a T beam as the width of the web (or rib) plus one-fifth of the distance between points of zero moment, and of an L beam as the width of the web plus one-tenth of the distance between points of zero moment. In both cases an upper limit is placed on the width in that the effective width must not exceed the actual width of the flange. Comparison of these recommendations with those given in Chapter 2 as being relevant to elastic design suggests some discrepancy between the two approaches. Direct comparison is difficult, however, and since the compressive stresses are rarely critical in design, is of academic interest only; BS 8110 brings the recommendations more into line with current thinking.

In theory, two cases can arise, the neutral axis of the section being either in the flange or in the web of the section. In practice, however, the latter condition is not common.

The design equations for the case where the NA lies within the flange of the section are formulated in exactly the same way as those for the singly reinforced rectangular beam. Referring to Fig. 3.16, moments about the line of action of the tensile forces give

$$M_u = 0.45 f_{cu} b_e 0.9x(d - 0.45x) \qquad (3.21)$$

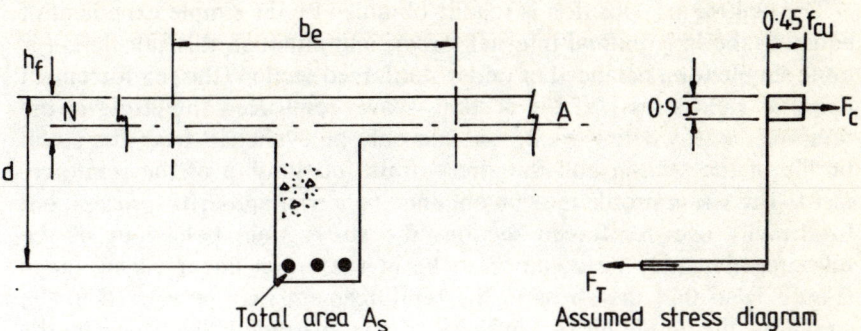

Fig. 3.16   T beam.

Moments about the line of action of the compressive force give

$$M_u = 0.87 f_y A_s(d - 0.45x) \qquad (3.22)$$

In the limiting condition, $x = h_f$, so that substituting into eqs. (3.21) and (3.22) gives the limiting condition for collapse moment as

$$M_u = 0.4 f_{cu} b_e h_f(d - 0.45h_f) \qquad (3.23)$$

or

$$M_u = 0.87 f_y A_s(d - 0.45h_f) \qquad (3.24)$$

Although these equations provide enough information to allow the design of a flanged section to be carried out, either of the design graphs previously obtained for the singly reinforced rectangular section may be used, with the effective width of the flange being used in place of the section width. The same procedure applies if the NA is in the web except that the longitudinal stresses in the web should be taken into consideration, so that the basic equations require slight modification.

3.7   Analysis of beam sections

The analysis of existing sections is no less important than the design of sections, but if the general principles of the design process are understood, the procedure of analysis readily follows. The main difference between design and analysis lies in the identification of the neutral axis position, which in the case of existing sections must be determined before the moment capacity of the section can be found.

The neutral axis position is readily obtained by the simple expedient of equating the longitudinal internal forces, and although this calculation is quite simple for a balanced or under-reinforced section (the reinforcement being at yield stress), if the section is over-reinforced the stress in the reinforcement is below yield, and can only be evaluated from the strain profile of the section and the stress-strain relationship of the reinforcement. The strain profile may be obtained by a trial-and-error process, but for heavily over-reinforced sections the stress-strain behaviour of the reinforcement may be assumed to be in the initial linear elastic range (Fig. 3.5) so that the stress in the reinforcement may be related to the strain by the modulus of elasticity of the material. The steps in the analysis procedure therefore are:

1. Determine an initial value of the neutral axis position by equating longitudinal forces (assuming the reinforcement to be at yield), i.e. $0{\cdot}4f_{cu}bx = 0{\cdot}87f_yA_s$.
2. If this value of x is equal to or less than $d/2$, the section is under-reinforced and the ultimate design moment is obtained from eq. (3.5). If x is greater than $d/2$, the section is over-reinforced and the estimate of x obtained from Step 1 will be too large, since the reinforcement will not be at yield stress. Determination of an accurate value of x must therefore take into account the strain profile of the section.
3. If the section is heavily over-reinforced (as indicated by an initial value of x greater than approximately d), the strain in the reinforcement may be assumed to be in the initial linear range. Referring to Fig. 3.17

$$\epsilon_s = \frac{d - x}{x}\,\epsilon_c$$

so that

$$f_s = E_s\epsilon_s = \frac{E_s\epsilon_c(d - x)}{x}$$

Hence the force in the reinforcement for design purposes is

$$F_t = \frac{E_s\epsilon_c(d - x)}{x}\,A_s\,\frac{1}{\gamma_m} \qquad (3.25)$$

and as $F_c = 0{\cdot}4f_{cu}bx$, equilibrium of the internal forces allows x to be evaluated. (Note that, once x is found, ϵ_s must be determined to ensure that the assumption that ϵ_s is in the initial linear part of the stress-strain curve is in fact true (i.e. $\epsilon_s < \epsilon_{sy}$).) If the section is not heavily over-reinforced, a trial-and-error procedure is required. A value of x is assumed and the corresponding reinforcement strain calculated from Fig. 3.17. Reference to the design stress-strain curve (as typified by Fig. 3.5) then allows the stress, and hence the force, in the reinforcement to be found. This is then compared with the compressive force in the section. If $F_c > F_t$, the assumed value of x is too large; conversely, if $F_c < F_t$, the assumed value is too small.

Since Step 1 overestimates the value of x for an over-reinforced section, the estimate of x for Step 3 should be reduced from that given by Step 1. A reasonable first approximation is obtained if x is taken as being midway between the value obtained from Step 1 and $d/2$.

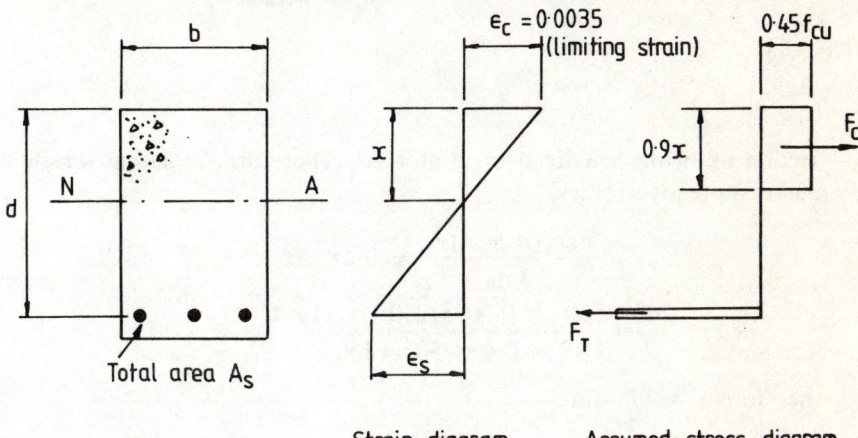

Strain diagram   Assumed stress diagram

Fig. 3.17 Analysis of singly reinforced section.

4. The ultimate design moment is then calculated from

$$M_u = F_t z \quad \text{or} \quad M_u = F_c z$$

Example 3.5 (analysis of existing section)
Determine the ultimate design moment of the section shown in Fig. 3.18 if $f_y = 460 \text{ N/mm}^2$ and $f_{cu} = 30 \text{ N/mm}^2$.

Following the procedure laid down above:

1. Initial value of x obtained is from

$$x = \frac{0 \cdot 87 f_y A_s}{0 \cdot 4 f_{cu} b} = \frac{0 \cdot 87 \times 460 \times 1964}{0 \cdot 4 \times 30 \times 150} = 437 \text{ mm}$$

2. $x > d/2$, therefore the section is over-reinforced. Since $x > d$, take the

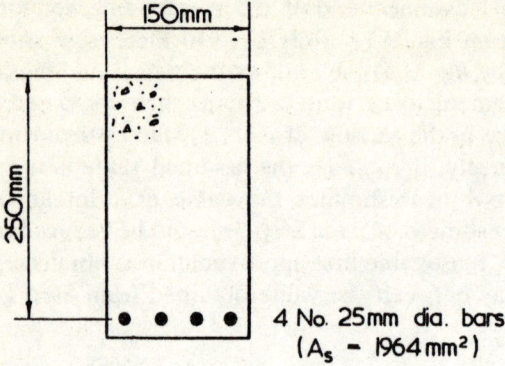

Fig. 3.18

section as being heavily over-reinforced. Therefore, equating tensile and compressive forces,

$$\frac{E_s \epsilon_c (d - x) A_s}{x \gamma_m} = 0 \cdot 4 f_{cu} bx$$

$$\frac{200 \times 10^3 \times 0 \cdot 0035 (250 - x) 1964}{1 \cdot 15 \times 0 \cdot 4 \times 30 \times 150} = x^2$$

therefore $x = 193$ mm.

Check that the assumption that ϵ_s is less than yield strain is true:

$$\epsilon_s = \frac{250 - 193}{193} \times 0 \cdot 0035 = 0 \cdot 001 < \epsilon_{sy}$$

hence the assumption is correct,

3. $$M_u = F_c z = 0 \cdot 4 \times 30 \times 150 \times 193 (250 - 87)$$
$$= 56 \cdot 6 \text{ kN m}$$

Reference to Section 3.3.1 shows that for design purposes a maximum value was placed on the depth to the neutral axis of $d/2$. Increasing the value of x beyond this therefore has no effect on the ultimate *design* moment of a section, which is limited to that given by eqs. (3.6) and (3.7). This simplifies the analysis of existing sections, since if the recommendations of BS 8110 are followed and Step 1 of the analysis shows the neutral axis depth to be greater than $d/2$, the ultimate moment of the section is calculated as if the value of x were $d/2$, and Step 3 of the analysis procedure is not required.

The analysis of doubly reinforced sections follows much the same process, except that determination of the position of the neutral axis requires that account be taken of the force in the compression reinforcement. It may not be assumed, however, that the compression reinforcement is at maximum stress since, even if $d'/d < 0.2$, if the depth to the neutral axis is very small the strain in the compression reinforcement may still be less than that giving maximum stress. For a complete analysis, therefore, a procedure that is similar to that used in Step 3, page 64, must be followed: a neutral axis position is assumed, and the strain in the compression reinforcement evaluated from the resulting strain profile. (This takes the limiting compressive strain in the concrete to be 0.0035.) From the stress-strain relationship (Fig. 3.5), the stress, and hence the force, in the compression reinforcement may be calculated, and equating the longitudinal forces then shows whether the assumed neutral axis position is correct. If not, the assumed value must be modified and the procedure repeated.

For most sections the depth to the neutral axis is sufficiently great that the compression reinforcement is at maximum stress. This is always so if the ratio d'/x does not exceed a certain value, which depends on the grade of reinforcement being used. For reinforcement having a yield stress of 460 N/mm^2 the limiting ratio is 0.43, whilst for reinforcement having a yield stress of 250 N/mm^2 the ratio rises to 0.7.

Analysis of doubly reinforced sections is illustrated by Examples 3.6–3.8.

Example 3.6
A rectangular beam section has dimensions and reinforcement as shown in Fig. 3.19. If $f_{cu} = 25$ N/mm^2, $f_y = 250$ N/mm^2 and cover to all reinforcement is 50 mm, determine the ultimate design moment of the section.

Equating the longitudinal forces,

$$F_c + F_{sc} = F_t$$
$$(0.4 \times 25 \times 250 \times x) + \left(628 \times \frac{250}{1.15}\right) = \frac{250 \times 3217}{1.15}$$

therefore

$$x = 225 \text{ mm}$$

x is therefore greater than $d/2 (d = 500 - 50 - 16)$, and the limiting value of x to be used is $d/2 = 217$ mm. The ultimate moment is given by eq. (3.15):

$$M_u = (0.155 \times 250 \times 434^2 \times 25) + [0.87 \times 628 \times 250 \times (434 - 50 - 10)]$$
$$= 233.5 \text{ kN m}$$

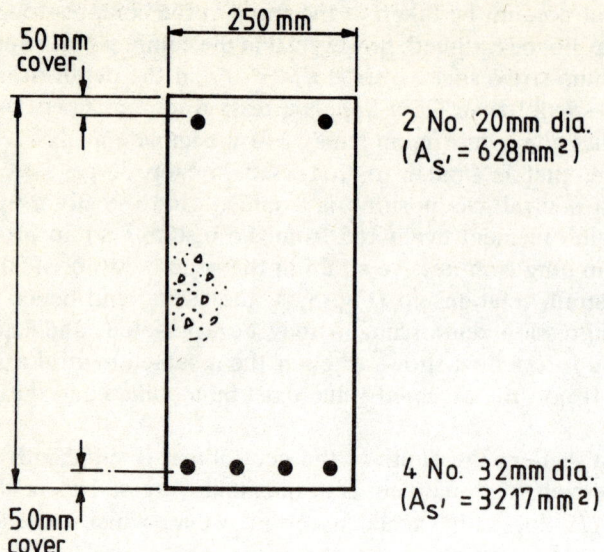

Fig. 3.19

Example 3.7

Repeat the analysis of the section shown in Fig. 3.19, but assume that the compression reinforcement is provided by four 20 mm diameter bars. All other conditions remain as before.

Equating the longitudinal forces gives

$$x = 170 \text{ mm}$$

Since $x < d/2$ and $d'/x \ngtr 0.4$, the actual value of x may be used. The ultimate moment is given by eq. (3.16):

$$M_u = [0.4 \times 25 \times 250 \times 170 \times (434 - 76.5)] + (0.87 \times 250 \times 1256 \times 374)$$
$$= 254 \text{ kN m}$$

Example 3.8

The analysis of Examples 3.6 and 3.7 is repeated, but with the compression reinforcement increased to four 25 mm diameter bars ($A'_s = 1964 \text{ mm}^2$) and with $f_{cu} = 35 \text{ N/mm}^2$. The strain profile of the section is as shown in Fig. 3.14, and the stress-strain relationship of the reinforcement is assumed to be as shown in Fig. 3.5.

Assume $x = 90$ mm, therefore the force in the concrete

$$0.45 \times 35 \times 250 \times 90 \times 0.9 = 319 \text{ kN}$$

From the strain diagram, the strain in the compression reinforcement = 0·0017, and from the stress-strain curve this produces a stress of 214 N/m². Therefore

$$F_{sc} = 214 \times 1964 = 420 \text{ kN}$$

and the total compressive force = 739 kN.

The total tensile force = $250 \times 0.87 \times 3217 = 700$ kN, hence, since $F_t < F_{sc} + F_c$, the assumed value of x is too large.

Try $x = 87$ mm: therefore $F_c = 308$ kN and $F_{sc} = 387$ kN, giving a total compressive force of 695 kN. The total tensile force is unchanged at 700 kN, so that a value of $x = 87$ mm may be assumed. From eq. (3.16),

$$M_u = [0.4 \times 250 \times 35 \times 87 \times (434 - 39)] + (387 \times 10^3 \times 371.5)$$
$$= 264 \text{ kN m}$$

Note that $d'/x = 62.5/87 = 0.72$. This is greater than the 0·7 ratio, showing that a detailed analysis is required.

3.8   Deflection

So far the discussion in this chapter has only considered one form of limit state, that of collapse (ultimate limit state). The consideration of deflection introduces a second limit state—serviceability. The serviceability limit state is not confined to deflection, but applies to any condition that will render the structure unsuited for its design purpose—hence serviceability would include vibration and cracking as well as deflection. The serviceability condition is covered in the British Standard by either using 'deemed to satisfy' conditions, such as limiting the beam span/depth ratio, or by a more basic analytical procedure that compares the calculated values of deflection etc. with accepted values. For most situations an analysis based on an elastic technique will be sufficient to provide a good indication of the deflection, but care should be exercised because the deflection response of a reinforced concrete section to increasing loads will not necessarily be linear as cracking developing throughout the section will change the effective section modulus, and creep effects will cause a variation in the Young's modulus of the material. However, in many instances, the technique outlined in Chapter 2 will provide a sufficiently accurate indication of the deflection to enable this condition to be assessed.

3.9   Compression members

Any discussion of compression members usually divides columns into the two categories of short and long columns, the division being intended to take account of the instability of long compression members. In the past, allowance for this instability effect was made by the provision of reduction factors which were intended to relate the permissible load on a long column to that for a short column of the same cross-section. The factors used were independent of the type of loading, however, so that both bending moment and axial load were effectively reduced. This is unrealistic since the primary difference between the behaviour of a slender column and that of a stocky column is due to the fact that the former experiences a greater lateral deflection, which produces an additional bending moment in the section; it is therefore unnecessary to reduce both the bending moment and the axial load.

The method now adopted is to consider the effect of slenderness as producing a potential increase in the eccentricity, so that a reduction of the moment capacity is achieved without a reduction of the axial load. The approach is somewhat complex in its analysis, but leads to the definition that a short or stocky column is one whose effective height is not greater than 10 times its least lateral dimension for an unbraced column, and 15 times for a braced column.

3.9.1   Columns under axial load

In short reinforced columns with little or no eccentricity of load the concrete always fails first, since the failure strain of steel in compression is greater than the failure strain of concrete. However, both materials carry their maximum stress, since the reinforcement is strained beyond yield.

The stress pattern assumed to act in the materials is as shown in Fig. 3.20. Since uniaxial compression is assumed, the concrete failure stress is taken as being $0.67f_{cu}$, so that the ultimate design stress is $(0.67/\gamma_m)f_{cu} = 0.45f_{cu}$. The compressive stress in the reinforcement is obtained from Fig. 3.5, from which the design stress in compression can be seen to be

$$f_y/\gamma_m = 0.87f_y.$$

Analysis of the section follows the usual procedure so that, using Fig. 3.20 and equating the forces in the concrete and the reinforcement to the applied load,

$$N = 0.45f_{cu}bh + 0.87A_{sc}f_y \qquad (3.26)$$

In practice, however, it is unlikely that true axial loads will be obtained, because of loading eccentricities produced by constructional tolerances. Allowance is made for this by reducing the stresses used in eq. (3.26) by approximately 10 per cent, this being considered sufficient to allow for an eccentricity of 0.05 times the minimum cross-sectional depth. Eccentricities above this level produce load situations that cannot be considered as axially loaded conditions. The equation giving the load in an axially loaded column therefore becomes

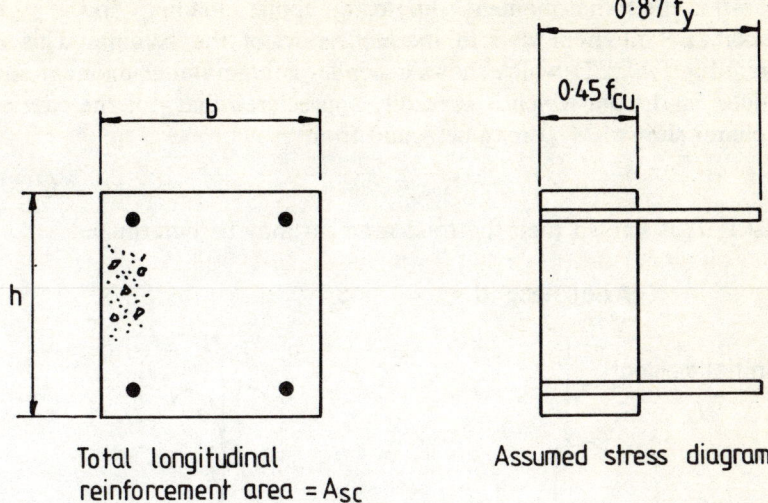

Fig. 3.20 Short, axially loaded column.

$$N = 0.4f_{cu}bh + 0.75A_{sc}f_y \qquad (3.27)$$

except in the case of columns supporting a symmetrical (or approximately so) arrangement of beams, when the ultimate load given by eq. (3.27) is reduced further to allow for any moments in the column arising from possible unsymmetrical loading of the beams. In this case the ultimate load reduces to

$$N = 0.35f_{cu}bh + 0.67A_{sc}f_y \qquad (3.28)$$

It will be noted that no allowance is made for the area of concrete displaced by the reinforcement.

The analysis of the behaviour of long columns under axial load is complicated by virtue of the additional moments that are introduced into the section by the deflection of the column under load. This is not purely a function of the length of the column, nor is it strictly accurate to refer to 'long columns' since the important parameter in this form of behaviour is the slenderness ratio of the column. This is the ratio of the length to the least lateral dimension, and it is this property rather than the length of the section that defines the behavioural characteristic. For a slender column the method of design is basically similar to that for a short column, except that the additional moment introduced by the column deflection should also be included. The total moment M_t in the column is given by

$$M_t = M_i + M_{add} \qquad (3.29)$$

where M_i is the initial moment (due to the applied loading) and M_{add} is the additional moment due to the deflection of the column. This is illustrated in Fig. 3.21 which shows a slender column under moment and axial load, and from which it is readily appreciated that, for the case of the column shown, M_{add} may be found from

$$M_{add} = Na_u \tag{3.30}$$

so that if a_u is known then the total moment may be determined.

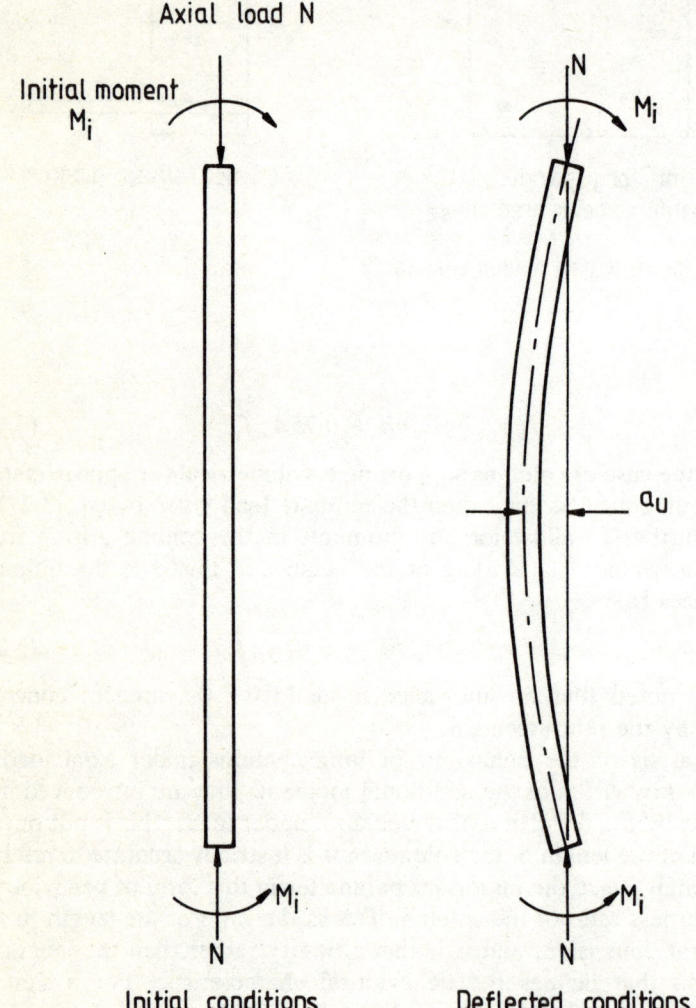

Fig. 3.21 Slender column under axial load and bending.

The value of the additional load eccentricity a_u due to the column deflection is dependent upon the curvature of the deflected column, which varies according to the initial shape of the column and the loading conditions. However, assuming that the failure is balanced (i.e. that the strains in both concrete and steel reach yield value), and that the curvature between the column ends and the mid-point is something between linear and uniformly varying (as shown in fig. 3.21), the value of a_u is obtained from

$$a_u = \beta_a h$$

where h is the depth of the cross section in the plane being considered,

and
$$\beta_a = \frac{1}{2000} \left(\frac{l_e}{b'}\right)^2$$

For uniaxial bending, b' is the smaller dimension of the column. For biaxial bending, however, additional moments will be introduced for each direction of bending, in which case b' should be taken as h, the dimension of the column in the plane of bending being considered.

Hence,
$$M_t = M_i + \frac{Nh}{2000} \left(\frac{l_e}{b'}\right)^2 \tag{3.31}$$

For the unbalanced type of failure, a reduction in the M_{add} value is made. This is achieved by multiplying M_{add} by a factor k which is dependent upon the ultimate axial load, the capacity of the column section under pure axial load and the axial load corresponding to the balanced condition, but in no case can the factor exceed unity.

3.9.2 Eccentrically loaded columns

The analysis of sections subjected to axial load and bending is complicated by the interaction of the axial and bending forces. This interaction may be illustrated diagrammatically, as in Fig. 3.22, from which it may be appreciated that three basic load conditions are pertinent: firstly the situation which produces axial stresses only, secondly the case of a 'small' bending moment, from which arises a condition of compressive stress only but of varying magnitude, and thirdly the condition of high bending moment and axial loads in which tensile stresses are developed in the section.

As with the loading conditions previously considered, the basic equations of equilibrium concern the equality of (a) longitudinal forces and (b) internal and external bending moments.

Suppose a column is under a load system which produces the third of the conditions listed above, i.e. a tensile stress exists in the section.

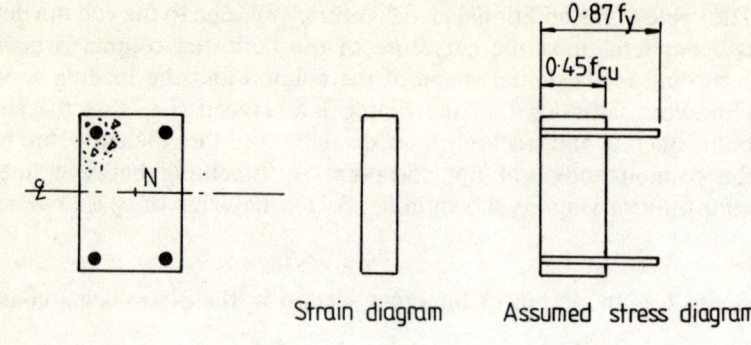

Strain diagram   Assumed stress diagram

(a) Axial loads only

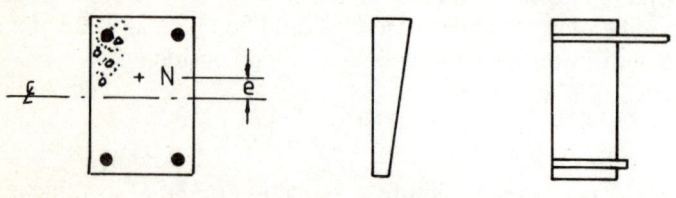

(b) "Small" bending moment (Ne)

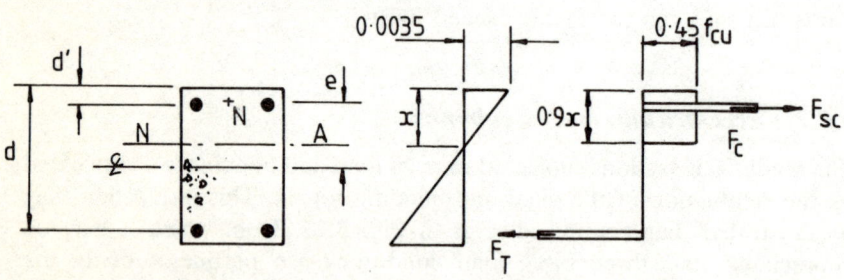

(c) "Large" bending moment (Ne)

Fig. 3.22 Interaction of bending and axial forces.

Figure 3.22(c) shows the stresses and forces that are developed. Considering the equilibrium of longitudinal forces,

$$N = F_c + F_{sc} - F_t \qquad (3.32)$$

and, taking moments about the neutral axis,

$$N\left(e - \left(\frac{d}{2} - x\right)\right) = F_c 0.55x + F_{sc}(x - d') + F_t(d - x) \qquad (3.33)$$

These two equations are perfectly valid in principle, but their solution produces some difficulty in practice. The forces in the reinforcement are directly related to the strains, and the assumption that plane sections remain plane (implying that the stress-distribution is linear) therefore fixes the position of the neutral axis and the depth of the compression block. If the strain-distribution is further assumed to be such that the reinforcement just reaches yield in both tension and compression, then the solution is comparatively simple, but this is rarely if ever obtained in practice. The actual strain distributions (and hence the position of the neutral axis) may be determined by a trial-and-error procedure, since, by assuming a value of x, values of N may be obtained from eqs. (3.32) and (3.33). The true value of x is that which produces the same value of N from both equations.

Although this procedure is suitable for analysis where the dimensions, reinforcement size, etc., are known, it is not suited to design as there are too many independent variables. The ultimate load design of columns under both axial and bending loads is therefore best performed with the aid of design graphs.

Consider the case of a rectangular reinforced concrete column having the dimensions and reinforcement details shown in Fig. 3.23. The strain and assumed stress-distribution diagrams are also shown, and a further assumption is that a balanced section exists, i.e. the concrete is crushing and the steel is yielding (in both compression and tension).

Equating longitudinal forces,

$$N = 0.4f_{cu}bx + 0.87A'_s f_y - 0.87A_s f_y \qquad (3.34)$$

From the strain diagram,

$$\frac{x}{0.0035} = \frac{(d - x)}{\epsilon_s}$$

but

$$\epsilon_s = \frac{f_y}{E_s}$$

therefore

$$\frac{x}{0.0035} = \frac{E_s(d - x)}{f_y} \qquad (3.35)$$

Assume $f_y = 460$ N/mm² and $E_s = 200 \times 10^3$ N/mm². This implies a high

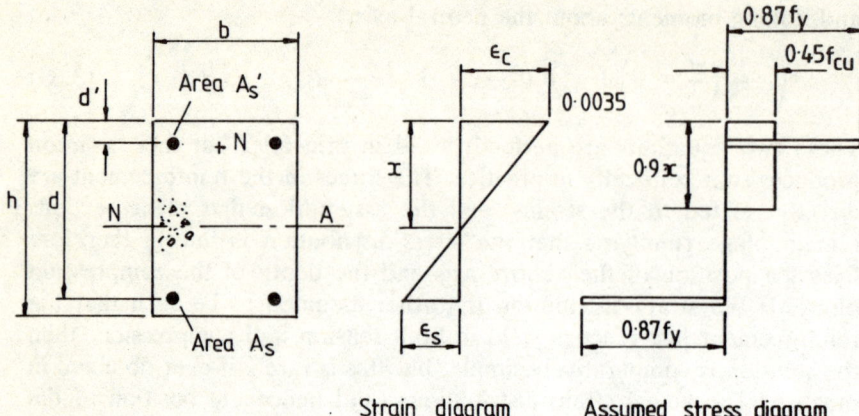

Strain diagram   Assumed stress diagram

Fig. 3.23 Rectangular column under bending and axial load.

yield steel reinforcement, but note that E_s will not change for a lower grade steel. Substituting into eq. (3.35) gives

$$x = 0.6d \qquad (3.36)$$

and putting this value of x into eq. (3.34),

$$N = 0.24f_{cu}bd + 0.87A'_sf_y - 0.87A_sf_y \qquad (3.37)$$

But load N is at an eccentricity e, thereby producing a moment in the column of Ne. The load system may therefore be considered as comprising an axial load N and a moment M_u, where $M_u = Ne$.

Taking moments about the tension reinforcement,

$$M_u + N\left(d - \frac{h}{2}\right) = 0.24f_{cu}bd[d - (0.45 \times 0.6d)] + 0.87A'_sf_y(d - d')$$

therefore

$$M_u + N\left(d - \frac{h}{2}\right) = 0.175f_{cu}bd^2 + 0.87A'_sf_y(d - d') \qquad (3.38)$$

Substituting into (3.37) for $f_y = 460$ N/mm², and dividing by bh,

$$\frac{N}{bh} = 0.24f_{cu}\frac{d}{h} + \frac{400A'_s - 400A_s}{bh} \qquad (3.39)$$

and assuming (as is the usual case) that $A_s = A'_s$, then

$$\frac{N}{bh} = 0.24f_{cu}\frac{d}{h} \qquad (3.40)$$

Substituting into (3.38) for N and f_y, remembering that $d' = h - d$ and rearranging, gives

$$\frac{M_u}{bh^2} = 0 \cdot 12 f_{cu} \frac{d}{h} - 0 \cdot 065 f_{cu} \left(\frac{d}{h}\right)^2 + \frac{200 A_{sc}}{bh} \left(\frac{2d}{h} - 1\right) \qquad (3.41)$$

where

$$A_{sc} = A_s + A_s'$$

so that for a given section in which f_{cu}, f_y, bh and A_{sc} are all known, N and M_u may be determined. This is for a *balanced* section. If the actual value of the ultimate load is greater than that found from eq. (3.37), the failure will be compressive. Thus, by considering eqs. (3.40) and (3.41) together with the ultimate load equation developed for axially loaded columns, the two extreme values N/bh and M_u/bh^2 for compressive failure may be found. For all conditions between these two extremes, failure will be compressive. This is shown in graphical form in Fig. 3.24, where position A represents the extreme of an axially loaded column ($M_u = 0$), and position B the other extreme of a balanced failure as indicated by eqs. (3.40) and (3.41). At this stage it is assumed that the relationship of N and M_u between these two points is linear.

If the eccentricity is greater than that which produces a balanced failure, then the tensile steel will yield before the concrete crushes, and in this case strain compatibility need not be considered in order to obtain N. Referring to eq. (3.34), and assuming that $A_s' = A_s$, allows eq. (3.42) to be obtained:

$$\frac{N}{bh} = 0 \cdot 4 f_{cu} \frac{x}{h} \qquad (3.42)$$

Taking moments about the tension reinforcement and resolving the load to be axial N plus a moment M_u,

$$M_u + N \left(d - \frac{h}{2}\right) = 0 \cdot 4 f_{cu} x b (d - 0 \cdot 45 x) + 0 \cdot 87 f_y A_s' (d - d') \quad (3.43)$$

for $f_y = 460$ N/mm^2 this leads to

$$\frac{M_u}{bh^2} + \frac{N}{bh} \left(\frac{d}{h} - \frac{1}{2}\right) = 0 \cdot 4 f_{cu} \frac{x}{h} \left(\frac{d}{h} - 0 \cdot 45 \frac{x}{h}\right) + \frac{200 A_{sc}}{bh} \left(\frac{2d}{h} - 1\right) \quad (3.44)$$

For given values of d/h, f_{cu} and A_{sc}, x/d for a certain N/bh may be evaluated from eq. (3.42). Substitution back into eq. (3.44) then gives M_u/bh^2, thus permitting the graph of N/bh against M_u/bh^2 to be drawn for the tension failure situation. This follows the part BC of the curve in Fig. 3.24. For any one set of values of f_{cu}, d/h and f_y, a family of curves may be drawn for various values of A_{sc}.

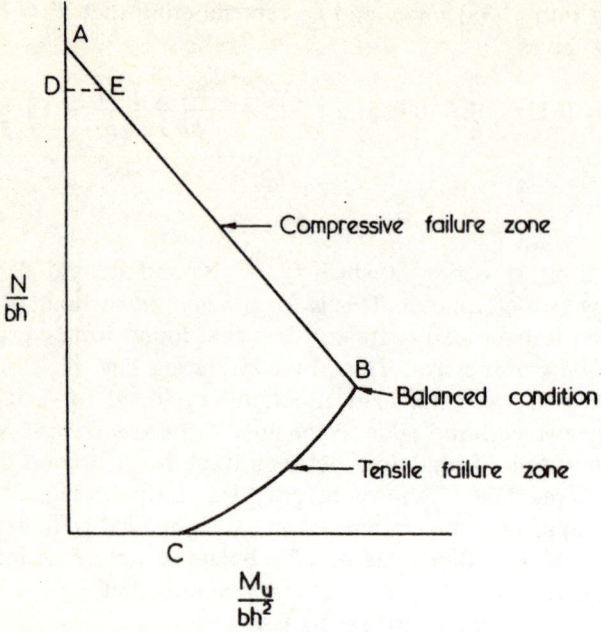

Fig. 3.24   Graph of N/bh against M_u/bh^2 for a rectangular column.

3.9.3 BS 8110 column design graphs

Part 3 of BS 8110 contains various design charts including a comprehensive selection of graphs to assist in the design of rectangular columns. The charts produced in the British Standard have been drawn up using the same basic procedure as outlined above, but the equations used differ slightly from equations 3.40 to 3.44 as the rectangular parabolic stress block for concrete in compression has been used in place of the simpler rectangular stress block used here. In addition, the BS 8110 graphs have required slight modification to smooth the curves in the region of low moments, where since the neutral axis lies outside of the section, part of the parabolic element of the stress block is also outside of the section and the standard equations therefore require some modification. A typical design graph from BS 8110 is presented as Fig. 3.25, from which it will be seen that a range of curves for varying A_{sc}/bh ratio are presented for each f_y, f_{cu}, and d/h ratio.

The use of the design graphs is very simple, and merely involves choosing a section size, calculating M_u/bh^2 and N/bh, and from the appropriate graph reading off the value of $100A_{sc}/bh$, from which the required reinforcement area may be calculated.

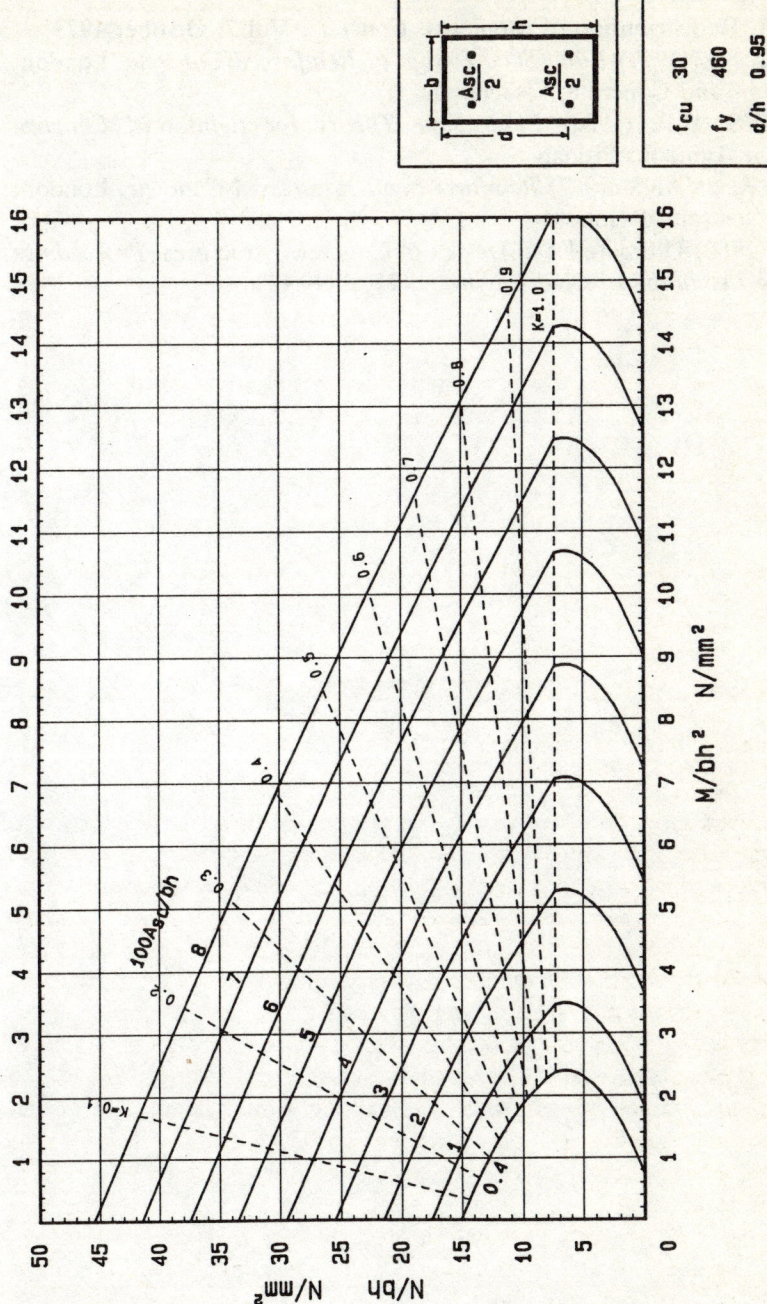

Fig. 3.25   Reproduced courtesy of BSI. Source: BS 8110 Part 3, Chart No. 30.

Additional reading

ASTILL Redistribution of Moments. *Concrete*, Vol 7, October 1973

BAKER, A. (1970) *Limit State Design of Reinforced Concrete*. London: Cement and Concrete Association

HUGHES, B.P. (1980) *Limit State Theory for Reinforced Concrete Design*. London: Pitman

KONG & EVANS (1987) *Reinforced and Prestressed Concrete*. London: Van Nostrand Reinhold

I.C.E. (1962) Ultimate Load Design of Concrete Structures. *Proceedings of the Institution of Civil Engineers*, **21**, 399–442

4 Slabs

4.1 Introduction

Slabs are probably the most commonly occurring structural elements, being used as floors, roofs, bridge decks, road and airfield pavements, and in foundation work. In its elementary form the slab is an extremely simple structure, but in practice the design/analysis is frequently complicated by the support conditions, by variation in the slab thickness and by the necessity of providing various openings in the slab. The depth of a slab is usually determined by deflection considerations rather than by the moment capacity, and the relevant codes of practice make recommendations of maximum span : depth ratios (i.e. minimum depth) to accommodate this. Limitation of the depth places a lower limit on the second moment of area value, so that as deflection is inversely proportional to the I value, the upper limit of deflection is obtained.

For short-span slabs the cheapest solution is to provide a slab of constant thickness over the complete span, since the savings in construction costs outweigh the extra costs of materials. As the span increases, however, the material costs become a progressively greater proportion of the overall costs, so that for slabs above about 4 m in span a cheaper solution is to reduce the self-weight of the structure by providing a hollow or voided slab, the voids being formed either through the centre of the slab (a hollow slab) or in the lower surface only.

4.2 One-way spanning slabs

As its name suggests, a one-way spanning slab spans in one direction only, i.e. the supports are along two opposed sides. The slab may therefore be considered as a wide shallow beam, but it is more convenient for design purposes to take the slab as being composed of a number of beams of unit width, spanning between the slab supports.

Examination of elastic plate theory shows that a strip of unit width within the plate has a higher flexural rigidity than a corresponding isolated beam, due to the fact that the material on either side of the strip resists the transverse strain. The flexural rigidity of the plate strip therefore becomes $EI(1 - v^2)$ as against EI for the equivalent isolated beam. For a reinforced concrete slab, however, this effect may be ignored, since Poisson's ratio for concrete is so low (commonly in the order of 0·15) that the v^2 term may be neglected. The effect may also be ignored for the reinforcement, since this is in the form not of a plate or slab but of individual bars; one-way spanning slabs may therefore be treated as beams, subject to the provision that distribution reinforcement is placed across the slab to ensure that the individual beams act as a homogeneous whole.

The actual design procedures may follow either the modular ratio or the load factor method (Chapters 2 and 3).

4.3 Two-way spanning slabs

If a slab is supported on more than two sides, or on two adjacent sides, then it can no longer be acting as a one-way spanning member, since some forces and moments must also be acting in the other direction(s). Such a slab is known as a two-way spanning slab, and the analysis is very much more complex than for a one-way spanning slab. The analysis may be performed by either elastic or collapse methods, both of which present complicated solutions, and the reader is recommended to specialist literature for detailed treatment of these analyses. In order that some understanding of the design method may be obtained, a simple introductory explanation is presented here.

4.4 Elementary theory of elastic plates

The elementary theory of elastic plates considers the equilibrium of a thin uniform plate under the action of various loads and reactions that are normal to the plane of the undeformed plate. An element of the plate, of size $\delta x \times \delta y$ and under a load of ω/unit area, is shown in Fig. 4.1, together with the forces and moments acting on it. The notation used is that commonly adopted for the analysis of plates and slabs, so that the single subscript indicates the plane on which the bending moment or force acts, e.g. M_x is the bending moment acting on the x plane (the plane perpendicular to the x axis). Double subscripts indicate a torsional moment, the first subscript indicating the plane on which the moment acts, the second indicating that the moment is a torsional moment. The sign convention used is that positive moments cause sagging curvature, positive shear forces act on the positive face in the positive z direction, and

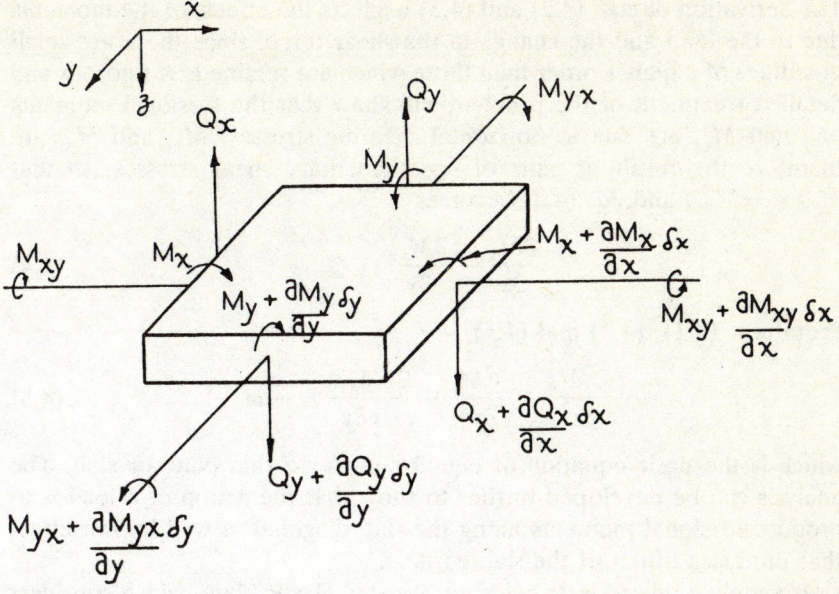

Fig. 4.1 Analysis of a plate.

positive torsional moments act in a clockwise direction about the outward-pointing normal.

Consider equilibrium of this element:

$$\omega\,\delta x\,\delta y + \left(Q_y + \frac{\partial Q_y}{\partial y}\,\delta y - Q_y\right)\delta x + \left(Q_x + \frac{\partial Q_x}{\partial x}\,\delta x - Q_x\right)\delta y = 0$$

therefore

$$\frac{\partial Q_x}{\partial x} + \frac{\partial Q_y}{\partial y} = -\omega \qquad (4.1)$$

Taking moments about the x axis and simplifying the equation,

$$\frac{\partial M_{xy}}{\partial x}\,\delta x\,\delta y - \frac{\partial M_y}{\partial y}\,\delta x\,\delta y + Q_y\,\delta x\,\delta y = 0$$

therefore

$$\frac{\partial M_{xy}}{\partial x} - \frac{\partial M_y}{\partial y} = -Q_y \qquad (4.2)$$

Similarly, for moments about the y axis

$$\frac{\partial M_{yx}}{\partial y} + \frac{\partial M_x}{\partial x} = Q_x \qquad (4.3)$$

The derivation of eqs. (4.2) and (4.3) neglects the effects of the moments due to the load and the change in the shear force, since these are small quantities of a higher order than those which are retained. A rigorous and detailed treatment of the problem will show that the torsional moments M_{xy} and M_{yx} are due to horizontal shearing stresses. M_{xy} and M_{yx} are therefore the result of pairs of complementary shear stresses, so that $M_{xy} = -M_{yx}$, and eq. (4.3) becomes

$$\frac{\partial M_x}{\partial x} - \frac{\partial M_{xy}}{\partial y} = Q_x \qquad (4.4)$$

From eqs. (4.1), (4.2) and (4.4),

$$\frac{\partial^2 M_x}{\partial x^2} + \frac{\partial^2 M_y}{\partial y^2} - \frac{2\partial^2 M_{xy}}{\partial x \partial y} = -\omega \qquad (4.5)$$

which is the basic equation of equilibrium for a thin plate or slab. The analysis can be developed further to show that the action of a load is to produce torsional moments along the slab diagonal, a well-known effect that produces lifting of the slab corners.

A complete theoretical analysis of the thin elastic plate, which considers both the equilibrium and compatibility equations together with the elastic stress-strain relationships, produces a fourth-order partial differential equation, the solution of which is extremely tedious for most practical shapes and support conditions. Semi-empirical methods are therefore used to enable analyses to be carried out readily. A criticism of such methods is that no allowance is made for the loss of flexural rigidity obtained by the cracking of the concrete in the tensile zones, but as this fact is also ignored in the theoretical analysis it is not in itself a justification for rejection of empirical methods.

The method used depends on the support conditions: for a simply supported slab the procedure adopted is that known as the Rankine-Grashof method.

4.4.1 Rankine-Grashof method

The Rankine-Grashof method assumes that the load on the slab is shared between strips of unit width running in the two directions parallel to the sides of the slab, as shown in Fig. 4.2. Because this method is restricted to slabs that are simply supported only, the corners are able to lift. This means that the torsional moments are zero, so that eq. (4.5) becomes

$$\frac{\partial^2 M_x}{\partial x^2} + \frac{\partial^2 M_y}{\partial y^2} = -\omega = -(\omega_x + \omega_y) \qquad (4.6)$$

By ignoring the interaction between the x and y direction strips, the equation reduces to the basic beam equations

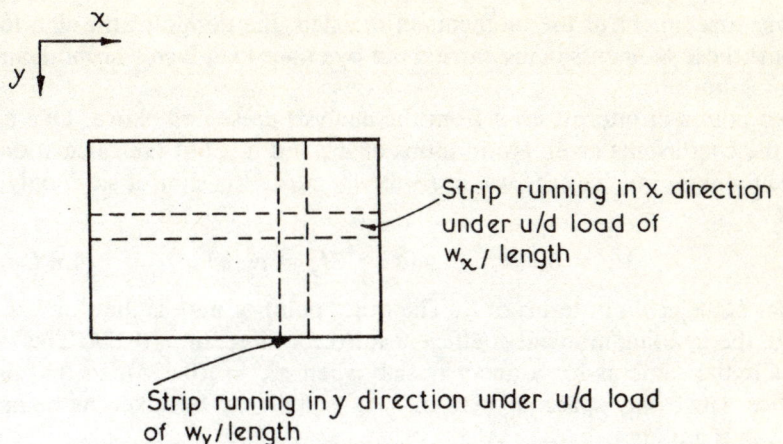

Fig. 4.2 Simply supported rectangular slab.

$$\frac{\mathrm{d}^2 M_x}{\mathrm{d}x^2} = -\omega_x \quad \text{and} \quad \frac{\mathrm{d}^2 M_y}{\mathrm{d}y^2} = -\omega_y \tag{4.7}$$

The strips may therefore be considered to act as beams and may be designed accordingly. The values of ω_x and ω_y are obtained from compatibility of equal deflections in the x and y direction strips at the centre of the slab. Hence, at the centre of the slab, $\Delta_x = \Delta_y$ but, since for the simply supported uniformly distributed load condition $\Delta = 5\omega l^4/384EI$,

$$\frac{\omega_x}{\omega_y} = \left(\frac{l_y}{l_x}\right)^4$$

In addition, $\omega_x + \omega_y = \omega$, so that it is now possible to write ω_x and ω_y in terms of the applied load ω and the lengths of the sides of the slab. For a simply supported beam under a uniformly distributed load, $M = \omega l^2/8$, so that the moments in the slab now become

$$M_x = \alpha_{sx} \omega l_x^2 \quad \text{and} \quad M_y = \alpha_{sy} \omega l_x^2 \tag{4.8}$$

where α_{sx} and α_{sy} are coefficients that are dependent upon the slab dimensions and various constants. Calculation of these coefficients may be illustrated by considering the case of a square slab, whence $l_x = l_y$. From the above, $\omega_x/\omega_y = 1$, so that $\omega_x = \omega_y = \omega/2$. Therefore

$$M_x = \frac{\omega_x l_x^2}{8} = \frac{\omega l_x^2}{16} = 0.062\omega l_x^2$$

i.e. $\alpha_{sx} = 0.062$, and in this case, by reason of symmetry, α_{sy} also $= 0.062$.

A similar procedure may be used for various ratios of l_x and l_y; the results are tabulated in BS 8110 and reproduced here as Table 4.1. This permits assessment of the moments in the slab, the design of the slab to support these moments being carried out by either load factor or modular ratio methods.

Two points of interest arise from the analysis presented above. One is that the coefficients given are in terms of α_{sx} and α_{sy}, but the calculation of the moment involves the use of l_x (the length of the shorter side) only. Thus

$$M_x = \alpha_{sx}\,\omega l_x^2 \quad \text{and} \quad M_y = \alpha_{sy}\,\omega l_x^2 \qquad (4.8 \text{ bis})$$

so that M_y is given in terms of l_x. The other point of note is that for l_y/l_x of 3·0, the bending-moment coefficient for the x direction is 0·124. This is virtually the same as for a one-way slab when $\alpha_{sx} = \frac{1}{8}(0\cdot125)$, so that in practice a two-way spanning slab of $l_y/l_x \geqslant 3\cdot0$ may be taken as being one-way spanning.

The design procedure using this method therefore becomes:

1. Evaluate l_y/l_x (l_x is in the shorter direction).
2. Evaluate α_{sx} and α_{sy} from Table 4.1.
3. Evaluate n (the total design ultimate load on the slab), from $n = 1\cdot4 \times$ dead load $+ 1\cdot6 \times$ live load.
4. Using $M_x = \alpha_{sx}nl_x^2$ and $M_y = \alpha_{sy}nl_x^2$, obtain the bending moments in the slab. (Note that both M_x and M_y use l_x to evaluate the moment.)
5. Design as for a simply supported beam under the moment given by Step 3.

Example 4.1

A concrete slab is 5 m long and 3 m wide, and is simply supported on all four sides. Determine the bending moments in the x and y directions if the live load is taken as being 700 N/m^2, the finishes dead load is 300 N/m^2 and the density of concrete is 2450 kg/m^3.

Assume a slab depth of 150 mm. The total loading then becomes

	N/m^2
live load	700
finishes	300
self-weight	3605

hence
$$G_k = 3905 \text{ N/m}^2, \quad Q_k = 700 \text{ N/m}^2,$$

and
$$n = 1\cdot4 \times 3905 + 1\cdot6 \times 700 = 6587 \text{ N/m}^2.$$

$l_y/l_x = 5/3 = 1\cdot66$, so that from table 4·1 $\alpha_{sx} = 0\cdot110$ and $\alpha_{sy} = 0\cdot04$. From eqs. (4.8),

$$M_x = \alpha_{sx}nl_x^2 = 0.11 \times 6587 \times 3^2 = 6521 \text{ N m}$$
$$M_y = \alpha_{sy}nl_x^2 = 0.04 \times 6587 \times 3^2 = 2371 \text{ N m}$$

Table 4.1 Bending-moment coefficients for slabs spanning in two directions at right angles, and simply supported on all four sides.

l_y/l_x	1·0	1·1	1·2	1·3	1·4	1·5	1·75	2·0	2·5	3·0
α_{sx}	0·062	0·074	0·084	0·093	0·099	0·104	0·113	0·118	0·122	0·124
α_{sy}	0·062	0·061	0·059	0·055	0·051	0·046	0·037	0·029	0·020	0·014

(Reproduced courtesy of BSI. Source: BS 8110 Part 1, Table 3.14)

The preceding analysis assumes that the slab is simply supported, and that the corners of the slab lift (i.e. the torsional moments are zero). If there is any restraint or fixity in the supports, the corners are obviously prevented from lifting, so that torsional moments are present and an analysis allowing for the built-in condition of support is required.

Although the elastic approach can be used in a semi-empirical form for this condition, it is much more convenient to use one of the collapse analysis techniques.

4.5 Collapse analysis of slabs

The methods of slab analysis so far considered are based on the elastic approach, although the design of the slab may utilise ultimate load procedures. It has long been realised, however, that although elastic methods offer a reasonable indication of structural behaviour at working loads, no accurate indication of the collapse load of the structure is obtained; as with the analysis of frames, plastic or collapse methods of analysis have been developed for slabs. The two principal methods are Johansen's yield line theory and the Hillerborg strip theory, both of which have the advantage over elastic analysis that they may be applied to more complex slab shapes. Although a detailed treatment of the methods is rather complex, the general procedures may be outlined as follows.

4.5.1 Yield line theory

The basis of the theory is that the moment-rotation curve of the slab has an idealised form as shown in Fig. 4.3. This assumption is quite reasonable because the slab depth is usually determined by deflection limits rather than by moment capacity, so that slabs are invariably under-reinforced.

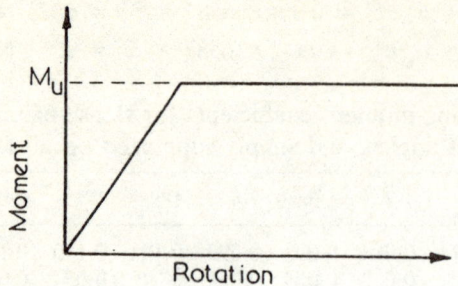

Fig. 4.3 Idealised moment-rotation curve.

The application of increasing load to the section therefore causes a corresponding increase in the bending moment until the yield moment is reached. Further increase of the load causes rotation of the section without increase in bending moment, the extra load being carried by the redistribution of the internal moments.

The development of the yield moment, together with the subsequent rotation, produces a plastic hinge which is signified by a yield line, but collapse will not occur until the moment redistribution has permitted enough yield moments to be developed to produce a mechanism. In the case of the slab, this means that all of the yield lines must reach the slab boundary for collapse to occur. The elements between the yield lines are subjected to elastic curvatures only, but these are very much less in magnitude than the plastic rotations at the yield surfaces; the elements may therefore be treated as rigid bodies, which means that the yield lines must be straight. The collapse mechanism of yield lines having been developed, the collapse load may be calculated by equating the internal and external work done in and on the slab.

One of the main disadvantages of the yield line approach is that it is necessary to assume the collapse mechanism. If the wrong mechanism is assumed, the collapse load will be too high, since the correct mechanism is always that which produces the lowest load. Although 'standard' solutions are available for a variety of slab shapes, it is frequently necessary to consider more than one yield line pattern. There are various conditions that must be fulfilled for a satisfactory yield line pattern to be developed, however, and these help in determining the more obvious mechanisms which usually predict collapse loads that are within 10 per cent of the lowest theoretical collapse load. Since the various approximations made in the theory can lead to a gain in strength of the experimental over the theoretical collapse load of about the same amount, the method provides a reasonable analysis of the actual performance of the slab.

The conditions aiding the prediction of the yield line pattern are:

1. Yield lines terminate at a boundary.
2. Yield lines are normally straight.
3. A yield line or yield line produced passes through the intersection of the axes of rotation of adjacent slab elements.
4. Axes of rotation usually lie along lines of support and pass over columns.
5. Symmetry of the slab and the loading applied produces symmetry of the yield line pattern.

The application of the method is illustrated by two simple examples. The first concerns a rectangular slab spanning between two simple supports, and subjected to a uniformly distributed load of ω/unit area. This problem is obviously too simple to justify the use of the yield line technique in a practical situation, and is included only to illustrate the basic procedure.

Example 4.2
The yield line pattern is assumed to be as shown in Fig. 4.4. This therefore means that the corresponding deflection may be assumed to be as shown in Fig. 4.4(b), with the yield line being displaced a distance δ.

The external work done is given by (load × distance moved) and, since a 'straight line' displacement is assumed, this may be evaluated as

$$\omega \times \frac{\delta}{2} \times \alpha L \times L = \frac{\omega \delta \alpha L^2}{2}$$

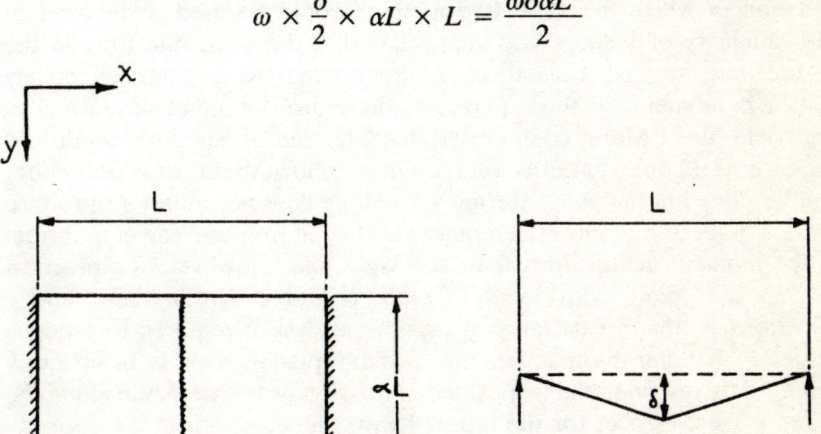

Fig. 4.4 Simply supported, one-way spanning slab.

The internal work done is due to the moments acting along the yield lines only, since, as twisting moments and shear forces on either side of a yield line are equal and opposite, so the internal work done *over the whole slab* due to these moments and forces is zero. The internal work is therefore given by (moment × rotation), where the moment is the moment of resistance of the yield line. This may most conveniently be expressed in terms of the resistance moment per unit length of yield line m, in which case the total moment is $m\alpha L$. The total rotation of the yield line is 2θ, so that the internal work done is

$$2\theta m\alpha L = m\alpha L \frac{2\delta}{L/2}$$

Equating the external and internal work,

$$\frac{m\alpha L4}{L} = \frac{\omega\delta\alpha L^2}{2}$$

therefore

$$m = \frac{\omega L^2}{8} \tag{4.9}$$

This is the same result as would be obtained by simple elastic analysis, since the failure mechanism is formed without any redistribution of moments. In practical terms this means that, once the first part of the slab has yielded, the whole yield line immediately forms.

The above example is, of course, extremely simple and is not the usual situation in which the yield line method would be used. One facet of the simplicity of loading and support is that the yield line runs in the y direction, so that evaluation of the internal work done effectively considers moments in the x plane, as these are the moments normal to the yield line. More complicated loading and/or support conditions produce yield line patterns that do not follow the x or y direction; equilibrating internal and external work done does not therefore produce a value of m_x or m_y, since the moment evaluated from such an equilibrium is the moment acting normal to the yield line. However, for practical reasons it is more convenient to place reinforcement in the x and y directions, so that for design purposes the analysis is required to produce values of bending moment in the x and y planes. This is most easily achieved by resolving the work done in rotation of the yield line along the x and y axes. Hence, for the internal work done,

$$\text{internal work} = \Sigma M\theta = \Sigma(M_y\theta_x + M_x\theta_y) \tag{4.10}$$

where M_x and M_y are the moments acting in the x and y planes and θ_x and θ_y are the components of rotation of the yield line about the x and y axes. M_x and M_y may equally well be written as $m_x l_y$ and $m_y l_x$, in which case m_x and m_y are the collapse moments of the slab in the appropriate planes, and l_x and l_y are the projected lengths of the yield line on the x and y axes. Then

$$\text{internal work} = \Sigma(m_y l_x \theta_x + m_x l_y \theta_y) \qquad (4.11)$$

The application of this approach is shown in Examples 4.3 and 4.4. The examples are still very simple and are not good illustrations of situations in which the yield line approach is best used, but serve to demonstrate the basic technique.

Example 4.3
Consider a square slab, simply supported on all four sides and having a load of ω/unit area uniformly applied over the total area (Fig. 4.5).

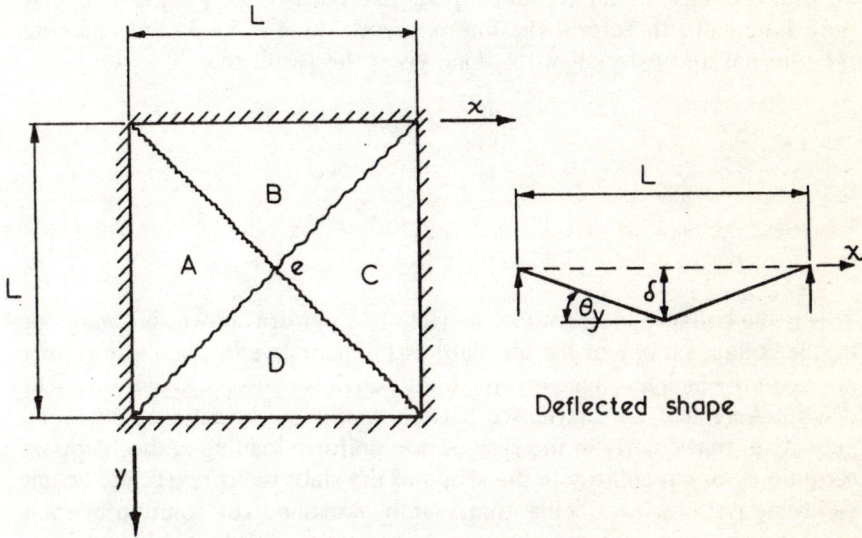

Fig. 4.5 Square slab simply supported on all four sides.

The yield line pattern is assumed to be that shown (note that the conditions for determining the yield line pattern as listed in section 4.5.1 are all fulfilled), and the centre of the slab e is assumed to deflect a distance δ.

The external work done per element is $[(\omega/2)L \times (L/2)]\delta/3$, where $\omega L^2/4$ is the total load on the element and $\delta/3$ is the average distance moved. (Notice that in this example the average distance moved is $\delta/3$, whereas in the previous case it was $\delta/2$. This is due to the differing shapes of the elements, which are triangular in this case but were rectangular in the previous example.)

The total external work done on the slab is therefore

$$4 \times \frac{\omega L^2 \delta}{12} = \frac{\omega L^2 \delta}{3} \qquad (4.12)$$

The internal work done by the slab is $\Sigma M\theta = \Sigma (M_y \theta_x + M_x \theta_y)$ where $M_x = m_x l_y$ and $M_y = l_x m_y$. l_x and l_y are the projections of the yield line on the x and y axes, so that, considering the element A, $l_x = L/2$ and $l_y = L$. θ_x and θ_y are the rotations of the yield lines about the x and y axes, respectively. For element A, θ_x is obviously zero and, from Fig. 4.5,

$$\theta_y = \frac{\delta}{L/2} = \frac{2\delta}{L}$$

The internal work done in rotation of element A is therefore $m_x 2\delta$, where m_x is the collapse moment in the x plane. Assume that $m_x = m_y = m$. For the whole slab, therefore, the internal work done is $8m\delta$, and equating the internal and external work done gives the result that

$$\frac{\omega L^2 \delta}{3} = 8\delta m$$

i.e.

$$m = \frac{\omega L^2}{24} \qquad (4.13)$$

This is the collapse moment for the yield line pattern shown, but may not be the collapse moment for the slab, as the actual yield line pattern may not be that which has been assumed. Although for this particular problem it is unlikely that an alternative pattern would be developed, situations may arise, particularly in the case of non-uniform loading and/or support conditions, or irregularity in the shape of the slab, that prevent the actual yield line pattern from being immediately assessed. The solution of such problems involves the consideration of several different patterns, the actual collapse moment being obtained from the pattern that predicts the lowest ultimate moment.

Suppose that, instead of being a square slab, the slab of Example 4.3 is rectangular, having one side of length L and the other of length αL. A yield line pattern following that for the square slab (as in Fig. 4.6(a)) might be obtained, but the change in side ratio does not make the slab doubly symmetric, so that the calculation used for the previous example must be slightly modified: instead of performing an evaluation of the work done on one element and then multiplying by 4 to give the total work done, the evaluation must be performed for two adjacent elements (say A and B) and symmetry then used to allow the total work done to be calculated.

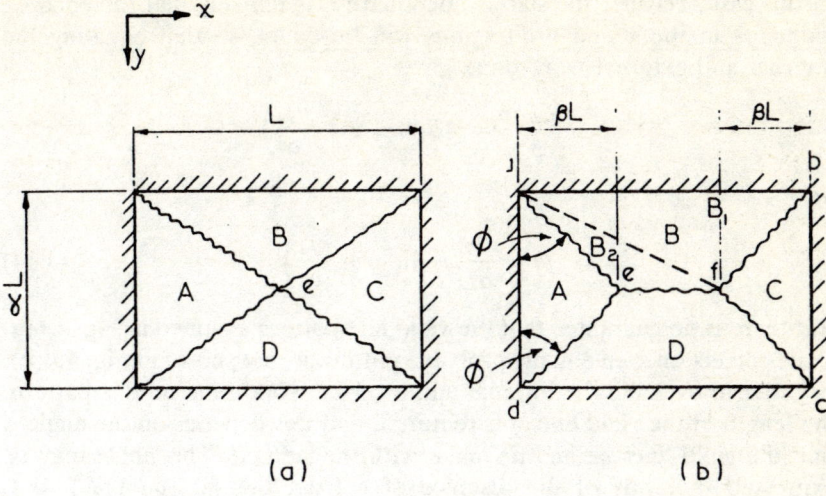

Fig. 4.6 Rectangular slab, simply supported on all four sides.

Suppose that, as before, the centre of the slab deflects a distance δ. For the yield pattern shown in Fig. 4.6(a), the external work done on element B is

$$\left(\tfrac{1}{2}\omega L\,\frac{\alpha L}{2}\right)\frac{\delta}{3} = \frac{\omega\alpha L^2\delta}{12}$$

and on element A is

$$\left(\tfrac{1}{2}\omega\alpha L\,\frac{L}{2}\right)\frac{\delta}{3} = \frac{\omega\alpha L^2\delta}{12}$$

The total external work done is therefore $(\omega\alpha L^2\delta)/3$.

Internal work done on element A is

$$m_x\,\alpha L\,\frac{2\delta}{L} = 2m_x\,\alpha\delta$$

and on element B is

$$m_y\,\frac{L2\delta}{\alpha L} = \frac{2\delta m_y}{\alpha}$$

so that for the whole slab the internal work done is

$$4m_x\,\alpha\delta + \frac{4\delta m_y}{\alpha} = 4\delta\left(m_x\,\alpha + \frac{m_y}{\alpha}\right)$$

In this case, because the slab is not square it is unlikely that the collapse moments in the x and y directions will be equal, so that equating the internal and external work done gives

$$\frac{\omega\alpha L^2\delta}{3} = 4\delta\left(m_x\alpha + \frac{m_y}{\alpha}\right)$$

or

$$\omega = \frac{12}{\alpha L^2}\left(m_x\alpha + \frac{m_y}{\alpha}\right) \tag{4.14}$$

But there is no guarantee that the yield line pattern assumed in Fig. 4.6(a) is the correct one; an equally probable pattern is that shown in Fig. 4.6(b). In order to evaluate the internal and external work done on this pattern, the length of the yield line ef is required, and this depends on the angle ϕ that the yield lines ae and de make with the side ad. This angle may be expressed in terms of the distance of e from the side ad (βL), but, whichever relationship is used; the work evaluations and hence the collapse moments are presented as a function of the variables ϕ or β.
Hence

$$m = \omega F(\alpha, \beta, \phi, \text{etc.})$$

or rewriting,

$$\omega = mF(\alpha, \beta, \phi, \text{etc.})$$

where F implies some function of the terms inside the brackets. The minimum value of ω is obtained by differentiation, so that for minimum ω, $d\omega/[dF(\ )] = 0$ and hence

$$\frac{\partial\omega}{\partial\alpha} = \frac{\partial\omega}{\partial\beta} = \frac{\partial\omega}{\partial\phi} = \ldots = 0 \tag{4.15}$$

For the slab and failure pattern shown in Fig. 4.6(b), evaluation of the internal and external work done follows a procedure similar to that for the previous pattern, in that the values are obtained for elements A and B, and symmetry then used to give the value for the complete slab.

The external work done on element A is

$$\omega\alpha L\,\frac{\beta L}{2}\,\frac{\delta}{3} = \frac{\omega\alpha\beta L^2\delta}{6}$$

Element B is most easily treated by subdividing into B_1 and B_2. The external work done on B_1 is

$$\omega L\,\frac{\alpha L}{2}\,\frac{1}{2}\,\frac{\delta}{3} = \frac{\omega\alpha L^2\delta}{12}$$

and on B_2 is

$$L(1 - 2\beta)\,\omega\,\frac{2\delta}{3}\,\frac{\alpha L}{2}\,\frac{1}{2} = \frac{\omega\alpha L^2\delta}{6}\,(1 - 2\beta)$$

So that the total external work done on the slab is

$$(2WDA + 2WDB_1 + 2WDB_2) = \omega\alpha L^2\delta[\tfrac{1}{2} - (\beta/3)] \qquad (4.16)$$

The total internal work is also obtained by considering elements A and B. For element A, $WD = m_x\alpha L\theta_y$, where $\theta_y = \delta/\beta L$. Therefore internal work done on element $A = m_x\,\alpha\delta/\beta$ and the internal work done on element $B = 2m_y\,\delta/\alpha$, so that the total internal work done is

$$(2WDA + 2WDB) = 2m_x\,\frac{\delta\alpha}{\beta} + 4m_y\,\frac{\delta}{\alpha} \qquad (4.17)$$

Equating the external and internal work gives

$$\omega = \frac{12m_y}{\alpha^2 L^2}\left[\frac{\alpha^2(m_x/m_y) + 2\beta}{3\beta - 2\beta^2}\right] \qquad (4.18)$$

This indicates that for a given slab in which α, L, m_x and m_y are all constant, ω is a function of β, so that for ω to be a minimum, $d\omega/d\beta = 0$. But

$$\frac{d\omega}{d\beta} = \frac{4\beta^2 + 4\alpha^2(m_x/m_y)\beta - 3\alpha^2(m_x/m_y)}{(3\beta - 2\beta^2)^2}$$

so that, if $d\omega/d\beta = 0$,

$$\beta = \frac{-4\alpha^2\,(m_x/m_y) \pm \sqrt{[16\alpha^4(m_x/m_y)^2 + 48\alpha^2(m_x/m_y)]}}{8} \qquad (4.19)$$

This value of β may then be substituted into eq. (4.18) to permit the collapse load for this yield line pattern to be determined. Leaving the solution in algebraic terms, however, produces an extremely unwieldy equation, and it is far more convenient to introduce the various slab dimensions and to continue the calculation in arithmetical terms.

Before going on to consider a worked example of a two-way spanning simply supported slab, it is of interest to examine the equations for ω obtained for the slabs so far considered. It will be seen that in each of the equations the deflection term δ is missing, and it is a general rule that equating the internal and external works eliminates δ from the load equation. It is not therefore necessary to know the actual displacement of the slab, so a displacement value of unity may conveniently be assumed.

Example 4.4
Returning to the simply supported two-way spanning slab, suppose that the slab under consideration measures 5 m $\times$ 3 m, has an effective

depth of 0·18 m and is reinforced by 8 mm bars at 150 mm centres (335 mm^2/m) running in the y direction, and 8 mm bars at 200 mm spacings ($251·5$ mm^2/m) running in the x direction. An f_{cu} of 25 N/mm^2 and f_{yt} of 460 N/mm^2 are assumed.

Referring to Fig. 4.6, the load causing collapse in the pattern of (a) is given by eq. (4.14), while that giving collapse in the pattern (b) is provided by eqs. (4.18) and (4.19). Both of these solutions require knowledge of m_x and m_y, which are the collapse moments of the slab in the x and y directions. These collapse moments are determined from the slab thickness and the reinforcement area by the methods set out in Chapter 3.

Equating the tensile and compressive forces for a one metre width of slab in the x direction gives

$$0·9x \times 0·45 \times 25 \times 1000 = \frac{251·5 \times 460}{1·15}$$

Therefore

$$x = 10 \text{ mm}$$

in the y direction:

$$0·9x \times 0·45 \times 25 \times 1000 = \frac{355 \times 460}{1·15}$$

therefore

$$x = 14 \text{ mm}$$

Assuming that the y direction reinforcement is placed on top of the x direction reinforcement (thereby reducing the y direction effective depth),

$$m_x = \frac{460}{1·15} \times 251 \times (180 - 4·5) = 17·6 \text{ kN m/m width}$$

and

$$m_y = \frac{460}{1·15} \times 335 \times (180 - 8 - 6) = 22·2 \text{ kN m/m width}$$

From the slab geometry, $\alpha = 0·6$ so that substituting into eq. (4.14) gives

$$\omega = \frac{12}{0·6 \times 25} \left(17·6 \times 0·6 + \frac{22·2}{0·6} \right)$$
$$= 38 \text{ kN/m}^2$$

This is for the yield line pattern as shown in Fig. 4.6(a).

From eq. (4.19),

$$\beta = 0.341$$

Substituting into eq. (4.18) gives

$$\omega = 36.2 \text{ kN/m}^2$$

which is the collapse load for the yield line pattern as shown in Fig. 4.6(b). In this particular case the collapse loads are very close, but indicate that the yield line pattern (b) is a more critical condition than pattern (a).

The examples so far considered are very simple, but serve to illustrate the basic procedures involved in the yield line analysis technique. More complicated problems involving edge fixity, concentrated loads, curtailment of reinforcement, openings in the slab or awkward slab shapes may be dealt with by the same basic procedure, but the equations derived are obviously more complex. A detailed treatment of the yield line method of analysis is too advanced to be considered at this stage. A list of further reading is included at the end of the chapter, but the brief introduction given above illustrates the value of the method.

4.5.2 Hillerborg's strip method

A variation of the collapse analysis of slabs, and one which has proved to be particularly efficient for the design procedure, is that known as Hillerborg's strip method. This assumes that the load at any point on a slab is transmitted to the nearest support point by bending in the plane perpendicular to that support. This implies that the slab may be treated as a number of strips, since a load applied at point X (Fig. 4.7), for example, will cause bending moments of m_x only about the support AB. The bending equation therefore reduces to the beam equation $d^2 m_x / dx^2 = -\omega$ (see eq. (4.7)), and the distribution of m_x may be calculated.

The form of this distribution is the same for both simply supported and

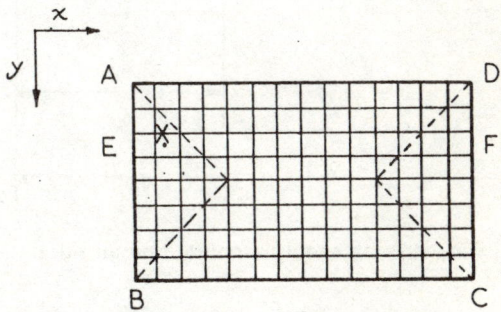

Fig. 4.7 Division of slab into strips (Hillerborg method).

fixed end supports, although the magnitude will of course vary. The strips running in the *x* direction (e.g. EF) will be loaded only at their ends, since the loads on the central portion will be carried on *y* direction strips out to the nearest support point. The distribution of moment along EF may therefore be obtained, so that a set of moment equations can be developed that is in equilibrium with the applied load. Reinforcement is provided to permit these moments to be carried by the slab, and this method thereby provides a design solution, whereas the yield line method provides an analytical solution.

The general technique may best be illustrated by considering the rectangular slab shown in Fig. 4.8. Assume that the slab is simply supported all round, with a uniformly distributed load of *w* applied over the whole area of the slab.

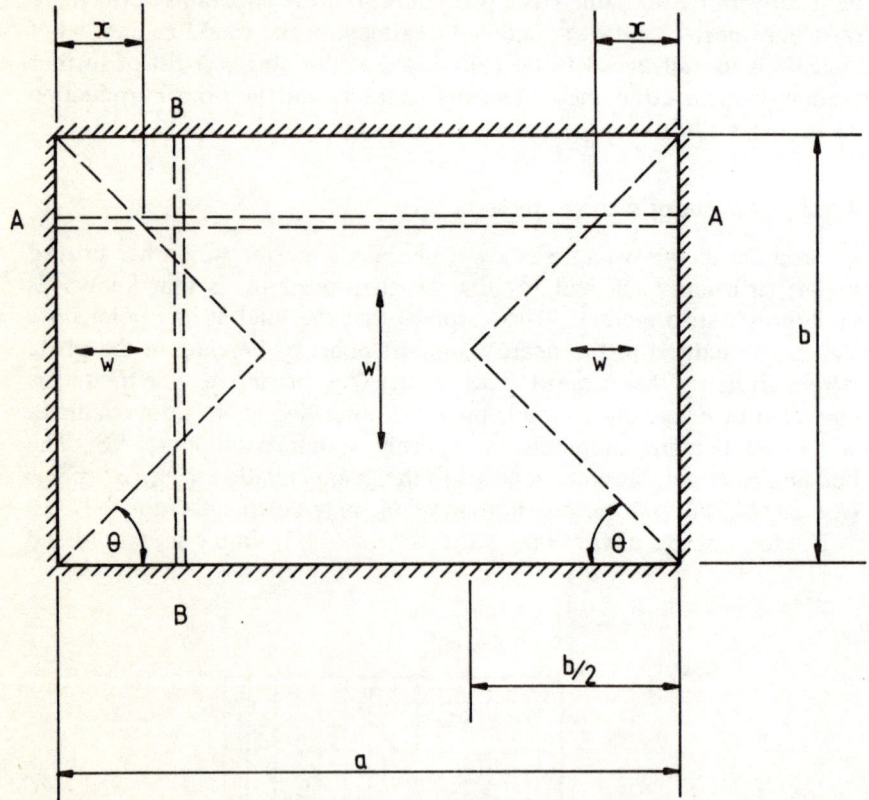

Fig. 4.8   Uniformly loaded slab simply supported on all sides.

Application of the basic principal of the load being transmitted to the nearest support point will give an effective load 'distribution' as shown.

Angle θ may be taken as being 45° for most situations, although a more detailed analysis would indicate that this is not the optimum angle of θ.

Consider the strip along $A-A$. This strip is spanning between the shorter sides of the slab, so that the ends of the strip are loaded by the total load that is being applied to the slab, whilst the central part has no load on it (all of the load being carried by strips spanning in the opposite direction, since the distance to the support is shorter). The loading and the bending moment diagrams are therefore as shown in Fig. 4.9

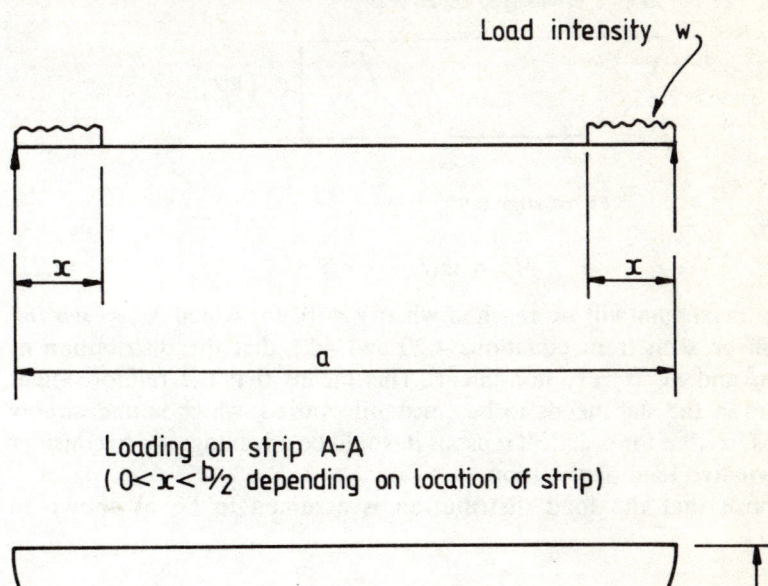

Load intensity w

Loading on strip A-A
($0<x<b/2$ depending on location of strip)

BM on strip A-A

Fig. 4.9

$$M_x = wx^2/2 \qquad (4.20)$$

and the maximum will be reached when $x = b/2$, in which case

$$M_x = wb^2/8$$

Consider strip $B-B$. The loading and bending moment diagrams are as shown in Fig. 4.10

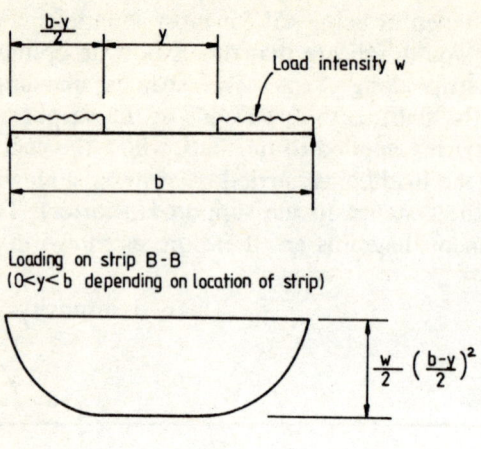

Fig. 4.10

$$M_y = w(b - y)^2/8 \qquad (4.21)$$

and the maximum will be reached when $y = 0$, for which $M_y = wb^2/8$

It will be seen from equations 4.20 and 4.21 that the distribution of both m_x and m_y is very non-linear. This means that the reinforcement provided in the slab needs to be constantly varied, which is undesirably complex, so that for practical reasons it would be advantageous to consider an alternative load distribution.

Suppose that the load distribution is assumed to be as shown in Fig. 4.11.

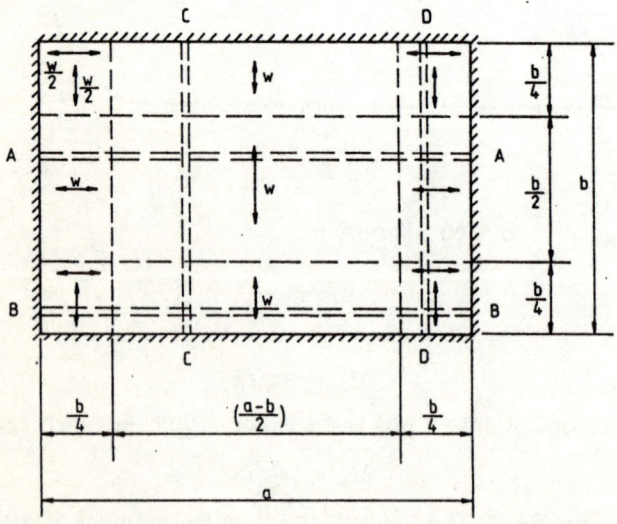

Fig. 4.11   Assumed load distribution for the slab from Fig. 4.8.

The division of the slab into middle and edge strips is made quite arbitrarily, except that it is advantageous to make the y direction edge strips the same width as the x direction edge strips. This leads to a simpler effective load distribution, particularly at the corners where for simplicity and ease of calculation, the basic assumption of all load being carried to the nearest support may be modified to give equal loads being carried in both directions. The assumed load distribution on the slab therefore becomes as shown.

Considering a strip along $A-A$

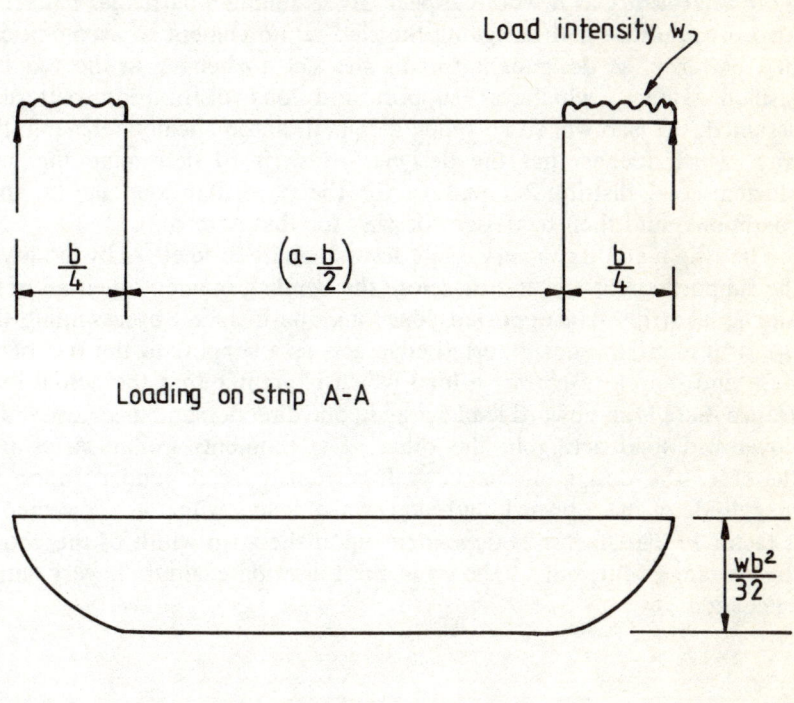

Load intensity w

$\frac{b}{4}$ $\left(a-\frac{b}{2}\right)$ $\frac{b}{4}$

Loading on strip A-A

$\frac{wb^2}{32}$

BM on strip A-A

Fig. 4.12

whilst for a strip along $B-B$, the maximum moment can be shown to be half of this value, the load intensity being $w/2$. A similar treatment for the y direction strips $C-C$ and $D-D$ shows that the bending moment distribution over $C-C$ is parabolic, leading to a maximum value of $wb^2/8$, whilst for the edge strip $D-D$ a distribution similar to that shown above is found, with the maximum moment being $wb^2/64$.

The load arrangement shown in Fig. 4.11 obviously provides a moment distribution throughout the slab that is much more uniform and regular than that arising from the load distribution assumed in Fig. 4.8, and which is therefore more suited to the design and construction procedures. Although the bending moments that are found in the slab are, on average, some 5 per cent higher than those obtained from the distribution shown in Fig. 4.8, the economies available from the more regular and simplified reinforcement pattern that is found outweigh this disadvantage, so that an assumed distribution similar to that shown in Fig. 4.11 is preferred.

The decision as to which pattern of load support is to be used is not quite as arbitrary as it would appear. By assuming a particular pattern of load distribution, and designing the slab reinforcement to accommodate that pattern, the designer is forcing the slab to behave in the required fashion. Hence, whichever support and load distribution pattern is assumed, the slab will (if correctly designed) automatically behave in that way, which means that the designer's task is to determine the most efficient load distribution pattern for the particular load and support conditions, and then to design the slab for that pattern.

The strip method can very easily accommodate fixed edges by modifying the support conditions and therefore the bending moment diagram of the individual strips. Unsupported edges may be included by assuming that the strip along the unsupported edge acts as a support to the rest of the slab, and therefore carries a load which is greater that the actual load. Hence there is an upward load acting in one direction and a corresponding downward load acting in the other. The moments in the strips (and therefore the design moments) will obviously be dependent upon the magnitude of the 'upward' and 'downward' loads, which are governed by a factor k. This factor is dependent upon the strip width of the chosen load arrangement, but as shown in the following example is very simply calculated.

Example 4.5

Figure 4.13 shows a rectangular slab that is simply supported along the two adjacent sides AD and DC, built in along a third side AB and unsupported along the fourth side. The slab carries a uniformly distributed load of intensity 10 kN/m^2 over the whole area, and it is required to determine the bending moments in the slab for the strip pattern shown.

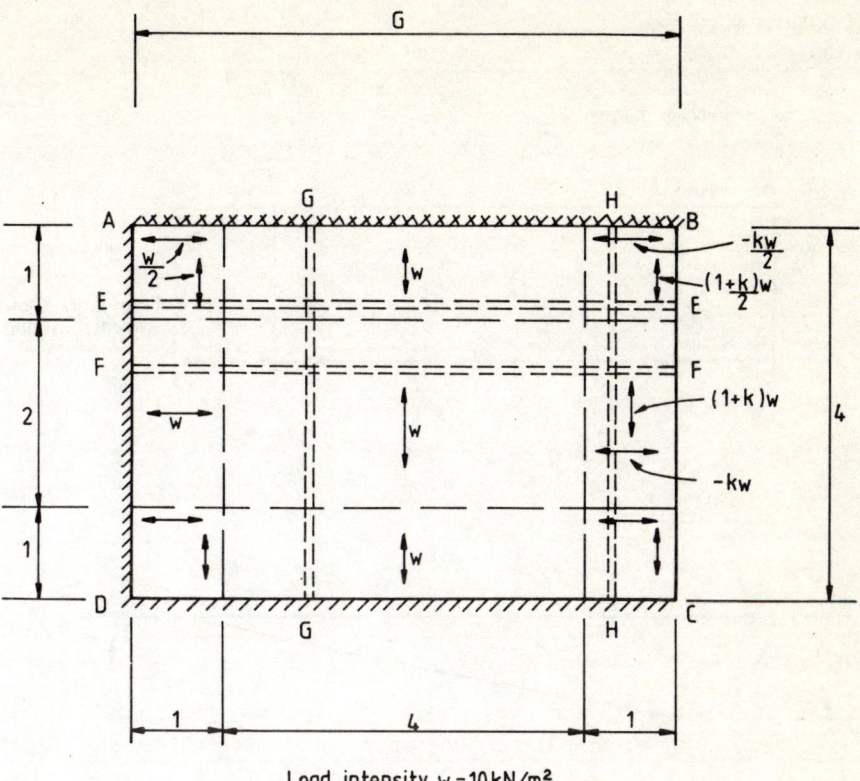

Load intensity w = 10 kN/m²

Fig. 4.13 Uniformly loaded slab with variable edge condition.

The load distribution obeys the general rule of the Hillerborg technique, namely that the load is transmitted to the nearest support, but is modified in the region of the unsupported side to take account of the imaginary support given by strip *HH*. This support is shown by the negative sign of the *k* factor, which implies an upward load on strips EE and FF, and which requires a corresponding increase in the downward load on HH in order to maintain the overall force balance.

Consider strip *EE*

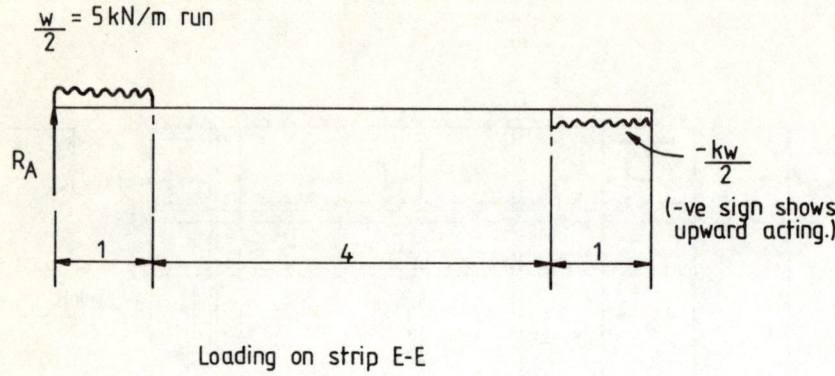

Loading on strip E-E

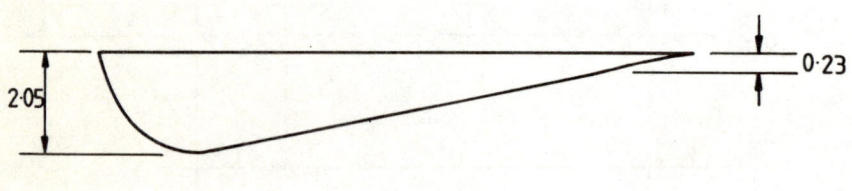

BM on strip E-E

Fig. 4.14

Simple analysis indicates that $k = 0.091$, that $R_A = 4.545$ kN, and leads to the bending moment diagram shown in Fig. 4.14. Strip *FF* has the same basic form of load and moment distribution, but since the applied load intensity is twice that of strip *EE*, the reaction and bending moments in strip *FF* are double those of strip *EE*. Note, however, that the value of k does not change.

Figure 4.15 illustrates the loading and support conditions of strip *GG*. Again the analysis is simple, although as the edge along *AB* is fixed, the system is statically indeterminate. However, as a standard load/support case, the moment diagram and the reaction values are easy to obtain.

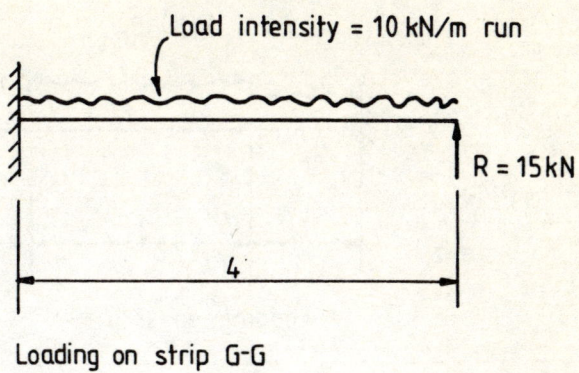

Loading on strip G-G

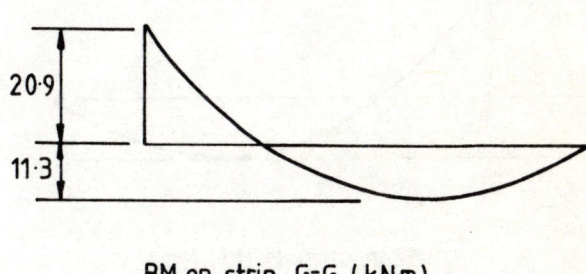

BM on strip G-G (kNm)

Fig. 4.15

Strip *HH* (Fig. 4.16) is rather more complex since it is also statically indeterminate, but has a non-uniform loading. Again, note that the value of k used is that obtained for the first strip (strip *EE*, Fig. 4.14).

The moments calculated from the above analysis are used as the design moments for the slab, and reinforcement provided accordingly, although consideration must obviously be given to the practicalities of the reinforcement sizes and spacings that result. Care should also be taken to ensure that sufficient attention is paid to the serviceability conditions, and that minimum reinforcement areas are provided.

The strip method can also accommodate variations in load intensity, slab shape etc. The essential feature of the technique, which must be remembered, however, is that as with the yield line approach, different arrangements of load produce different moments in the slab. However, unlike the yield line procedure, the strip method is inherently conservative, and therefore safe, although this may be outweighed by a some loss of economy.

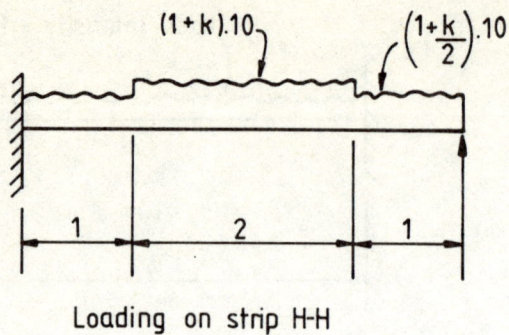

Loading on strip H-H

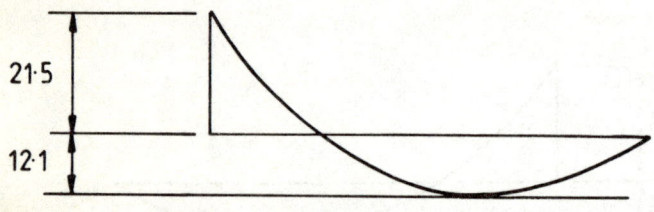

BM on strip H-H (kNm)

Fig. 4.16

The two methods outlined above are both well suited to the collapse analysis and design of slabs of both regular and irregular form, and solutions have been prepared for a number of the more commonly occurring slab formation and edge support conditions. These have been incorporated into BS 8110 in a tabular form with coefficients β_{sx} and β_{sy} being tabulated for rectangular slabs under various support conditions. These coefficients, which have been derived from the yield line approach, are used in exactly the same way as the coefficients for simply supported conditions that were presented in Section 4.4.1, and produce values of design moment in the x and y directions given by

$$m_x = \beta_{sx} w l_x^2$$
$$m_y = \beta_{sy} w l_x^2$$

Note that as in the simply supported example, both m_x and m_y are expressed in terms of the length of the shorter side l_x.

As with the earlier method, the coefficients are dependent upon the slab edge support conditions and on the ratios of the slab dimensions, and are restricted to regular rectangular slabs. Once the bending moments have been obtained, the reinforcement, slab thickness and other details are determined in the same way as is used in the design of beams.

4.6   Isotropic and orthotropic slabs

Slabs are frequently referred to as being either isotropic or orthotropic. These terms refer to the potential behaviour of the slab with regard to the ultimate or collapse flexural moments. If the design of the slab is such that the collapse flexural moments in two perpendicular directions in the plane of the slab are equal, then the slab is termed *isotropic*. If the collapse flexural moments in two perpendicular directions in the plane of the slab are different, the slab is termed *orthotropic*. The slab considered in Example 4.2 is therefore isotropic, since $m_x = m_y$, whereas the slab considered in Example 4.3 is orthotropic.

4.7   Shear

The shear behaviour of reinforced concrete sections has presented many problems to designers in the past and is still the subject of considerable discussion by research workers. In common with other areas of design, treatment of shear behaviour now follows the limit state philosophy, but the hitherto used elastic analysis provides a useful introductory background to the subject.

4.7.1   Elastic analysis of shear in a rectangular beam

Figure 4.17 shows a rectangular beam of width b carrying a bending moment M. Consider a small element of the beam, at a distance x_x from one end. A shear stress v acts on a horizontal plane LL at a distance y from the neutral axis. Considering the equilibrium of the prism $ABLL$

Compressive force acting from left to right

$$= \left(f_c + f_c \frac{y}{x} \right) \tfrac{1}{2}(x - y)b$$

Compressive force acting from right to left

$$= \left[(f_c + \delta f_c) + (f_c + \delta f_c) \frac{y}{x} \right] \tfrac{1}{2}(x - y)b$$

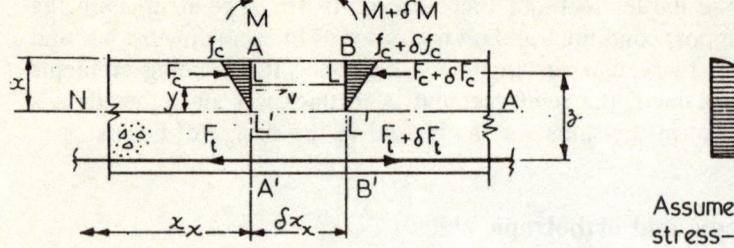

Assumed shear
stress—distribution

Fig. 4.17 Elastic analysis for shear.

Therefore the out-of-balance end force (obtained from the difference of the above two forces) is

$$F_{\text{int}} = \frac{\delta f_c b (x^2 - y^2)}{2x}$$

This is resisted by the shearing force acting on plane LL. Hence, for equilibrium,

$$v\delta x b = \frac{\delta f_c b (x^2 - y^2)}{2x}$$

therefore

$$v = \frac{\delta f_c (x^2 - y^2)}{\delta x \, 2x} \qquad (4.22)$$

This has parabolic variation with y, and reaches a maximum when $y = 0$.

Theoretically, the distribution curve for negative y (i.e. below the neutral axis) is of the same form, but in the case of a concrete section the material in the tensile zone is assumed to have cracked, so that no tensile forces may be carried by the concrete and the resultant horizontal force must be obtained from the reinforcement only. Hence if the prism $ABLL$ is transferred to the tension zone ($A'B'L'L'$), the out-of-balance end force is δF_t, and equating this to the shear force on plane $L'L'$ gives

$$v = \frac{\delta F_t}{\delta x b} \qquad (4.23)$$

There is therefore no variation of the horizontal shear force over the tensile zone of the section; it remains constant at its maximum value, except that no shear forces exist in the concrete below reinforcement level.

The force in the reinforcement may be equated to the geometry of the section and the applied moment, since equating the internal and external moments

$$F_t z = M$$

gives

$$\delta F_t = \frac{\delta M}{z}$$

so that

$$v = \frac{\delta M}{z \delta x b}$$

and putting

$$V = \frac{\delta M}{\delta x}$$

gives

$$v = \frac{V}{zb} \qquad (4.24)$$

which is the equation giving the shear stress in a section as used in codes based on elastic theory.

It is immediately obvious that this approach is an oversimplification of the problem, since the existence of shear stresses on plane $L'L'$ implies that complimentary shear stresses exist on planes $A'L'$ and $B'L'$. These faces are assumed to have cracked, however, so that no shear stresses can be developed on them, and thus theoretically the shear stress cannot be taken into the reinforcement. This obviously does not represent the true state of affairs, in which the shear forces are transmitted to the reinforcement by the elements or 'tongues' of concrete extending below the neutral axis between the tension cracks. The uniformity of shear stress predicted by the above analysis cannot therefore occur, particularly for that part of the section below the neutral axis, and some alternative analysis is required.

4.7.2 Collapse analysis of shear

The elastic analysis is based on an assumed behaviour which bears little resemblance to the true situation, whereas the collapse or limit state analysis attempts to consider the actual behaviour of the section. The analysis is of necessity somewhat complex, and the following account of the theory is considerably simplified.

Discussion of the limitations of the elastic approach to the problem of shear in reinforced concrete sections introduced the concept of the element or 'tongue' of concrete extending below the neutral axis, and bounded by flexure cracks. The effect of a shearing force on the section is to cause certain forces and moments to act on this element, the magnitude and combinations of which govern the shear resistance of the overall section. Analysis of the problem is complex, but the main contributions to the shear resistance come from the aggregate interlock forces along the line of the crack, the dowel action of the longitudinal reinforcement and the stresses in the uncracked compression zone. The relative magnitudes of these forces and the exact method by which they combine is imperfectly understood. Several authors have contributed to the knowledge of this subject; a detailed coverage of the area together with considerable background information may be obtained from Reagan and Yu (1973). However, the influence of the general approach may be clearly seen in the relevant BS 8110 clause, which now considers both reinforcement area and concrete strength in recommending the design shear stresses for a section. This contrasts with the elastic approach, in which the permissible shear stress is dependent upon the concrete grade only.

One of the main limitations of the theory is that it assumes that the shear stresses in the concrete tongues are not so high as to produce further cracking between the flexural cracks. This places an upper limit on the ultimate design shear stress (tabulated in BS 8110 and reproduced here as Table 4.2), which is effectively achieved by limiting the reinforcement ratio. The implication of this is that even if the reinforcement in the section is such that the value of $100\ A_s/bd > 3\cdot0$, the design concrete shear stress shown in Table 4.2 must not be exceeded.

Experimental evidence indicates that sections having an effective depth less than approximately 400 mm fail at a slightly higher shear stress than do sections of a greater thickness (assuming that all other parameters are equal), so that Table 4.2 presents values of design concrete shear stress for a range of effective depth values as well as $100\ A_s/bd$ ratios. In order to present a complex set of information in as simple a fashion as possible, the values of v_c given in Table 4.2 are for one grade of concrete only, having $f_{cu} = 25\ \text{N/mm}^2$. Other concrete grades require the design concrete stress obtained from Table 4.2 to be multiplied by $(f_{cu}/25)^{0\cdot33}$ as indicated in the note beneath the table.

The design shear stress at any section of a beam is given by

$$b_v = \frac{V}{bd} \qquad (4.25)$$

This equation is simple to use, but strictly speaking is only applicable to beams of uniform depth. If the design shear stress in the section exceeds the design concrete value given in Table 4.2, then shear reinforcement is required to increase the shear capacity of the section.

Table 4.2   Design concrete shear stress in beams (N/mm^2)

$\dfrac{100 A_s}{b_v d}$	Effective depth (in mm)							
	125	150	175	200	225	250	300	> 400
	N/mm^2	N/mm^2	N/mm^2	N/mm^2	N/mm^2	N/mm^2	N/mm^2	N/mm^2
≤ 0.15	0.45	0.43	0.41	0.40	0.39	0.38	0.36	0.34
0.25	0.53	0.51	0.49	0.47	0.46	0.45	0.43	0.40
0.50	0.67	0.64	0.62	0.60	0.58	0.56	0.54	0.50
0.75	0.77	0.73	0.71	0.68	0.66	0.65	0.62	0.57
1.00	0.84	0.81	0.78	0.75	0.73	0.71	0.68	0.63
1.50	0.97	0.92	0.89	0.86	0.83	0.81	0.78	0.72
2.00	1.06	1.02	0.98	0.95	0.92	0.89	0.86	0.80
≥ 3.00	1.22	1.16	1.12	1.08	1.05	1.02	0.98	0.91

NOTE 1. Allowance has been made in these figures for a γ_m of 1.25.

NOTE 2. The values in the table are derived from the expression:

$$0.79 \left(100 A_s/(b_v d)\right)^{1/3} (400/d)^{1/4}/\gamma_m$$

where

$\dfrac{100 A_s}{b_v d}$ should not be taken as greater than 3;

$\dfrac{400}{d}$ should not be taken as less than 1.

For characteristic concrete strengths greater than 25 N/mm^2, the values in table 3.9 may be multiplied by $(f_{cu}/25)^{1/3}$. The value of f_{cu} should not be taken as greater than 40.

(Reproduced courtesy of BSI. Source: BS 8110 Part 1, Table 3.9)

4.7.3   Shear reinforcement

If the design shear stress at a section is too high, then shear reinforcement is required. This is provided by means of either stirrups (links) or a combination of stirrups and bent-up bars, which carry only the difference between the design concrete and the design shear stresses.

The basis of the theory is the fact that pure shear acting on a section produces tensile and compressive stresses within the section. Consider the element shown in Fig. 4.18 which is assumed to be at the neutral axis of a concrete beam, and which has shear stresses acting on faces *AB* and *CD*.

For equilibrium, complementary shears of the same magnitude must act on faces *AD* and *BC* in such a direction as to prevent rotation of the element, so that the combined system becomes as shown.

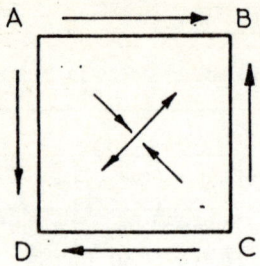

Fig. 4.18   Diagonal tension produced by shear.

Strength of materials theory shows that the effect of these stresses is to produce tensile and compressive stresses acting on the diagonal planes *BC* and *AD*. These are known as principal stresses, and are of the same magnitude as the original shear stresses. If the element of Fig. 4.18 is taken as being away from the neutral axis, then the magnitude of the principal stresses and the planes on which they act are affected by the flexural stresses in the beam, which of course are zero at the neutral axis. Application of this theory to concrete sections therefore suggests that the presence of shear stresses in a concrete section causes tensile cracking in the concrete at an angle of 45° to the neutral axis (i.e. perpendicular to plane *BD* of the element), and this leads to the theory known as the truss analogy.

This theory, which is one of the basic and long-standing theories relating to the shear behaviour of reinforced concrete sections, assumes that the reinforced section behaves in a fashion similar to that of a pin-jointed truss, with concrete taking the compressive forces and the reinforcement providing the tensile support. Although the analysis in its original form was quite simple, it produced a conservative solution as it made no allowance for the effects of the uncracked concrete in the compression zone, for the effects of aggregate interlock or for the dowel action of the longitudinal reinforcement; it presumed that, once the concrete had cracked as a result of the diagonal shear tension, no contribution could be expected from the concrete, and all tensile and shear forces would be carried by the reinforcement.

The results of a number of experimental investigations have indicated that the effects of the aggregate interlock, and other factors previously ignored, are significant. Although the truss analogy still forms the basis

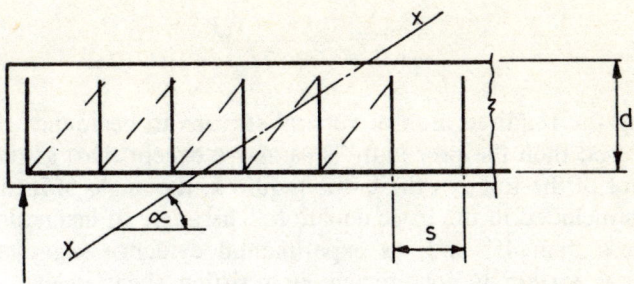

Fig. 4.19 Shear resistance of a reinforced section.

of the theory of shear behaviour of reinforced concrete sections, it is now modified to the extent that the shear strength of the concrete is assumed always to make some contribution to the overall strength of the section, even after cracking, and so allowance is made for the contributions of aggregate interlock, concrete in the compression zone and the dowel action of the main reinforcement.

Figure 4.19 shows a reinforced concrete beam with shear reinforcement provided by vertical stirrups. The stirrups and the main reinforcement represent the tensile members of a truss, and the concrete acts as the compression members. Even assuming that the concrete has cracked in diagonal tension, the aggregate interlock and other factors provide some shear resistance, v_c, in the concrete. At shear collapse, therefore, the external shear force on the section is resisted by the forces in the shear reinforcement plus the shear resistance of the cracked concrete, so that, considering the vertical equilibrium of the section to the left of the diagonal tension crack $X-X$,

$$vb_v d = \left(\frac{A_{sv} f_y}{\gamma_m} \times \frac{d}{s_v} \cot \alpha \right) + v_c b_v d \qquad (4.26)$$
$$\quad\quad\quad\;\; (1) \qquad\quad (2) \qquad\qquad (3)$$

where

$vb_v d$ = external shear
expression (1) = force in one stirrup
expression (2) = number of stirrups
expression (3) = shear resistance of cracked concrete

Note that the number of stirrups cut by the section line depends on the spacing of the stirrups, the depth of the beam and the angle of the tension crack which, for design purposes, may be taken as being 45°.

Rearranging eq. (4.26) and putting the partial safety factor for the shear reinforcement $\gamma_m = 1 \cdot 15$ gives

$$\frac{A_{sv}}{s_v} = \frac{b_v(v - v_c)}{0.87 f_y} \tag{4.27}$$

permitting the required area of vertical stirrups to be found. If inclined bars are used then the procedure is as above except that, as the vertical component of the force in the bar is required, the angle of inclination of the bar is included in the basic equation. The angle of inclination should not be less than 45° and, as experimental evidence suggests that an inclined bar system is not efficient in resisting shear unless additional reinforcement in the form of stirrups is provided, it is now recommended that a maximum of 50 per cent of the shear reinforcement be in the form of inclined bars, the remainder being vertical stirrups.

If the applied shear force is particularly high, there is a possibility that failure may occur in the concrete by crushing under diagonal compression. The likelihood of this type of failure is therefore eliminated by placing an upper limit on the shear stress acting on the section. This limit is dependent upon the concrete grade and is given by the lesser of $0.8\sqrt{f_{cu}}$ or 5 N/mm².

Another limitation applicable to the design of the shear reinforcement is that placed on the spacing of the stirrups. Too great a spacing may prevent the stirrups from crossing a shear crack, so a maximum spacing of $0.75d$ is specified.

Although the preceding discussion suggests that the shear reinforcement is only required if the actual shear stress exceeds the design concrete shear stress, it is possible for lightly reinforced beams to fail suddenly as a result of diagonal tension. In order to eliminate this, a minimum amount of shear reinforcement is recommended, although, if the actual shear stress is less than 0.5 of the design concrete value, even this nominal area may be omitted. The BS 8110 recommendation is that the minimum area should fulfil the requirements of eq. (4.28), and that this minimum area should be provided if the design shear stress lies between $0.5v_c$ and $(v_c + 0.4)$ N/mm². As with the main shear reinforcement, the stirrup spacing is limited to a maximum of $0.75d$.

$$\frac{A_{sv}}{s_v} \geq \frac{0.4 b_v}{0.87 f_{yv}} \tag{4.28}$$

For a design shear stress greater than $(v_c + 0.4)$ N/mm², shear reinforcement should be provided in accordance with eq. (4.27). The provision of a nominal area of stirrups has additional practical advantages, particularly in locating the beam reinforcement while placing the concrete, and it is good policy always to provide a nominal amount of shear reinforcement in the form of stirrups for purely practical reasons, even if no reinforcement at all is theoretically necessary.

Example 4.5

A rectangular reinforced concrete beam carries a uniformly distributed load of 80 kN/m over a simply supported span of 4 m. The beam is 0·25 m wide and 0·55 m deep, and is reinforced by four 25 mm diameter bars placed at an effective depth of 0·5 m. Assuming the concrete to be Grade 30, and that load factors of 1·6 on live loads and 1·4 on dead loads are applied, determine what shear reinforcement may be required.

The first step is to obtain the shear force diagram. This requires knowledge of the total load on the beam, i.e. the applied load plus the beam self-weight. Taking the density of reinforced concrete to be 2400 kg/m^3, the beam self-weight is

$$0·25 \times 0·55 \times 2400 \times 9·81 = 3·24 \text{ kN/m}$$

Therefore the total load on the beam is 83·24 kN/m.

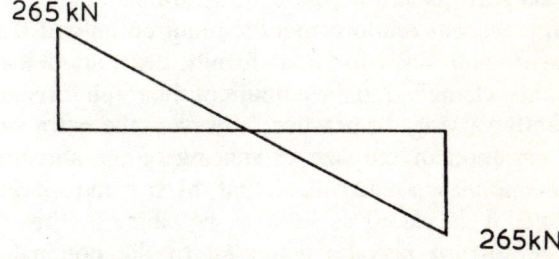

265 kN

265kN

Fig. 4.20 Shear force diagram for Example 4.5.

The shear force diagram is as shown in Fig. 4.20, from which it can be seen that the ultimate shear force is 265 kN. The beam dimensions must now be checked to ensure that the design shear stress does not exceed the maximum permissible.

$$\text{Design shear stress} = \frac{265 \times 10^3}{250 \times 500} = 2·12 \text{ N/mm}^2$$

From $v \ngtr 0·8\sqrt{f_{cu}}$, the maximum value for Grade 30 concrete is 4·4 N/mm^2, so that the beam dimensions are adequate. From eq. (4.25), $v = 2·12$ N/mm^2; if this is greater than the design concrete value given in Table 4.2, shear reinforcement is necessary.

$$\frac{100A_s}{b_v d} = \frac{100 \times 1964}{250 \times 500} = 1·57, \; \xi \, d = 500 \text{ mm}$$

Therefore, from Table 4.2,

$$v_c = 0·73 \text{ N/mm}^2$$

However, this is for a grade 25 concrete, so that in order to find v_c for the grade 30 concrete of this example

$$v_c = 0.73 \times \left(\frac{30}{25}\right)^{0.33}$$

Therefore

$$v_c = 0.78 \text{ N/mm}^2$$

Hence shear reinforcement is required, and is given by eq. (4.27).

Assuming that stirrups of 10 mm diameter mild steel are used ($f_{yt} = 250 \text{ N/mm}^2$), $A_{sv} = 157 \text{ mm}^2$. From eq. (4.27),

$$s_v = \frac{157 \times 0.87 \times 250}{250(2.12 - 0.78)} = 102 \text{ mm}$$

This spacing is less than the maximum permitted ($0.75d = 375$ mm), so that 10 mm dia stirrups at 100 mm centres are suitable.

This amount of shear reinforcement is required only at the position of maximum shear, and since for a uniformly distributed load the shear force is constantly changing, so the amount of shear reinforcement required is also constantly varying. In practice, however, the extra work involved in continual variation of the size or spacing of the stirrups more than nullifies any economies in material, so that the size and spacing of stirrups in any one section is varied as little as possible. In this example, the spacing of the stirrups may be increased to the nominal value when the shear stress in the section drops to the ultimate design value of 0.78 N/mm^2. This occurs when the shear force is 97.5 kN, i.e. at 1.26 m from each support.

For ease of construction this may be increased to 1.3 m. The provisions of eq. (4.28) apply over the central portion of the beam, and from this equation the nominal spacing of 10 mm dia stirrups required is

$$\frac{157 \times 0.87 \times 250}{0.4 \times 250} = 341 \text{ mm}$$

Therefore use 100 mm dia mild steel stirrups at 300 mm centres over the central 1.5 m of the beam, and 10 mm dia mild steel stirrups at 100 mm centres over the remainder of the span.

4.7.4 Shear in flat slabs

The problem of shear in slabs is dealt with in much the same way as that of shear in beams, except that two different forms of shear need to be considered. The first, which is exactly the same as that in beams, is the flexural shear, while the second is a more localised effect that is due to

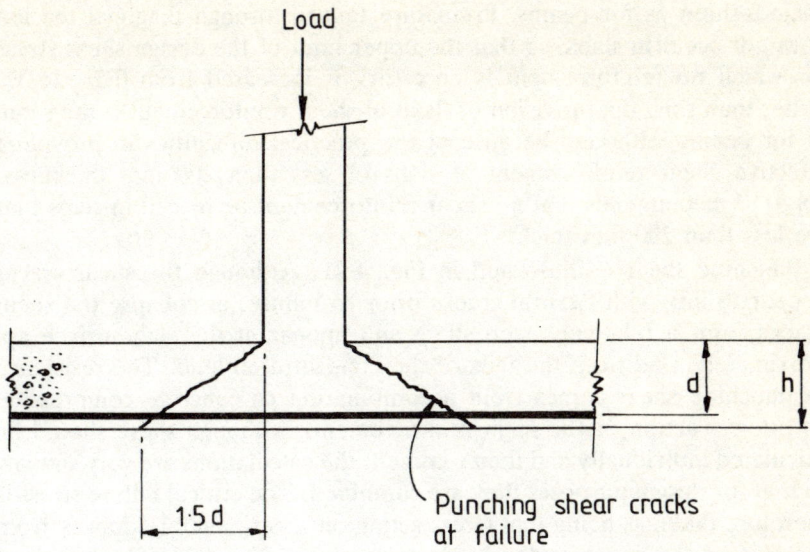

Fig. 4.21 Punching shear.

concentrated loads. The shear associated with this latter condition is frequently referred to as punching shear, and requires an approach somewhat different from that used for flexural shear.

For the flexural shear condition, the design concrete shear stress is obtained from Table 4.2, and shear reinforcement is provided in the same

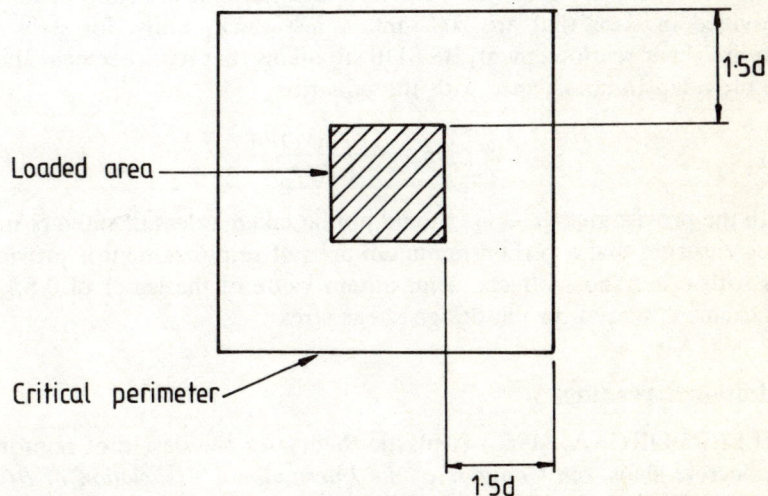

Fig. 4.22 Critical perimeter for punching shear.

basic fashion as for beams. Premature failure through diagonal tension does not occur in slabs, so that the upper limit of the design shear stress for which no reinforcement is necessary is increased from $0.5v_c$ to v_c. Other than this, the provision of flexural shear reinforcement is the same as for beams, although because of the practical difficulties in providing effective shear reinforcement in slabs of less than 200 mm thickness, BS 8110 recommends that no shear reinforcement be placed in slabs that are less than 200 mm thick.

Punching shear is illustrated in Fig. 4.21. Although the shear cracks appear to link with flexural cracks prior to failure, at collapse the shear cracks form a relatively even slope and appear at the slab surface approximately $1.5d$ from the face of the concentrated load. The resistance to punching shear comes from a combination of concrete compression and dowel action of the main reinforcement. Although these should be calculated individually and then summed, the calculations are very similar, so that for design purposes they are combined. The critical failure stress is therefore taken as being that stress acting on a perimeter $1.5d$ away from the face of the concentrated load, as shown in Fig. 4.22. The length of this perimeter is u, where

$$u = (\text{perimeter of concentrated load}) + 12d \qquad (4.29)$$

and the shear stress is assumed constant over the depth of the section.

The punching shear stress v obtained from this calculation is compared with the value of v_c appropriate to the section; if $v > v_c$, reinforcement is required. Provision of punching shear reinforcement is subject to the same practical limitations as apply to the provision of flexural shear reinforcement, however, so that no punching shear reinforcement is provided in slabs that are 200 mm or less thick, whilst for slabs that require shear reinforcement, BS 8110 stipulates that reinforcement should be provided in accordance with the equation

$$A_{sv} \sin \geq \frac{(v - v_c)ud}{0.87f_{yv}} \qquad (4.30)$$

with the proviso that $(v - v_c)$ should not be taken as less than 0.4 N/mm^2, thus ensuring that a certain minimum area of reinforcement is provided. As with other shear effects, a maximum value of the lesser of $0.8f_{cu}$ or 5 N/mm^2 is placed on the design shear stress.

Additional reading

HILLERBORG, A. (1960) A plastic theory for the design of reinforced concrete slabs. *6th Congress of the International Association of Bridge and Structural Engineers.*

HILLERBORG, A. (1974) *Strip Method of Design*. London: Viewpoint Publications, Cement and Concrete Association.

JOHANSEN, K.W. (1962) *Yield Line Theory*. London: Cement and Concrete Association.

JONES, L.L. and WOOD, R.H. (1967) *Yield Line Analysis of Slabs*. London: Thames & Hudson/Chatto & Windus.

KONG, F.K. and EVANS, R.H. (1987) *Reinforced and Prestressed Concrete*. London: Van Nostrand Reinhold.

REGAN, P.E. (1981) *Behaviour of reinforced concrete flat slabs*. London: CIRIA report no. 89.

REGAN, P.E. and YU, C.W. (1973) *Limit State Design of Structural Concrete*. London: Chatto & Windus.

WOOD, R.H. and ARMER, G.S.T. (1968) The theory of the strip method for design of slabs. *Proceedings of the Institution of Civil Engineers*, **41**, 313–331.

5 Prestressed Concrete Construction

5.1 Introduction

The dominant feature of all concrete design is the fact that, although strong in compression, concrete is so weak in tension that for design purposes the tensile strength may be taken as being effectively zero, and reinforcement must therefore be added to the section to carry tensile forces. The object of the prestressing technique is to introduce some degree of compressive stress into the 'tensile' zone before the flexural loads are applied, so that when the loads causing flexure come onto the section the resulting tensile stresses are reduced or even eliminated entirely. The whole of the concrete may thus be assumed to be effective, permitting a lighter construction for a given span. A disadvantage of the reduction of the tensile stresses is that the compressive stress in the section is increased, so that in order to use the prestressing technique a high quality concrete is required. The prestressing force is provided by high tensile rods or cables known as tendons, which are usually anchored to the ends of the concrete section, but which may be bonded to the concrete along part of their length.

Apart from allowing a lighter structure to be used, prestressing has several other significant advantages. The first of these is that, for most sections, the formation of flexural tension cracks at working moments is prevented, so that the effective depth of the section becomes the full depth. The avoidance of cracks gives the additional benefit of increased protection of the reinforcement against corrosion and fire; even if a portion of the concrete should crack as a result, perhaps, of overload, the presence of the prestressing force ensures that the tension cracks are closed if the overload is removed. Other advantages are that prestressing introduces a camber that is in opposition to the live load deflection and thus reduces the overall deflection; since diagonal stresses at working loads are reduced, less area is required to carry the shear forces on the section, so that T and I sections may be used, giving a further saving of material and dead weight.

There are two techniques of prestressing, known as pretensioning and post-tensioning.

5.1.1   Pretensioning

This technique involves stressing the tendons before the concrete is placed and maintaining the tension while the concrete sets. The tendons are then released and the reaction transferred into the concrete by bond stresses at the ends of the member in areas known as transmission zones. The full prestress is not effective in these areas, which should therefore be kept as short as possible.

Pretensioning is usually used in precast production line techniques, owing to the difficulty of obtaining suitable reactions for the initial tendon forces in in-situ work. For practical reasons it usually involves placing the tendons at a constant eccentricity throughout the length of the member.

5.1.2   Post-tensioning

This procedure applies the stresses to the tendons after the concrete has been placed in position and allowed to set. Ducts are placed in the formwork, and the concrete placed around these ducts. When the concrete has set, the prestressing wires are passed through the ducts, and the tendons tensioned by jacking against the ends of the section. There is no necessity to place the tendons after the concrete has been poured, and the ducts may be placed in position with the tendons already in place. After the prestressing force has been applied, the tendons are anchored by nuts or wedges which transfer the reaction to the concrete. The space between the tendon and the duct is then filled with grout (a cement/water slurry), which bonds the tendon to the duct (and hence to the member) throughout its length, thus ensuring composite action and protecting the steel from corrosion.

The post-tensioning procedure is entirely self-contained, with no requirement for external reactions to develop the prestressing force in the tendons, and is therefore very suitable for in-situ construction. A further advantage is that the position of the tendons can be altered along the section, thus varying the prestress moment.

5.1.3   Stresses in a prestressed member

The way in which the stress combinations are built up in a prestressed member is shown in Fig. 5.1, with the load conditions producing the various stresses described below. Although the signs of the stresses

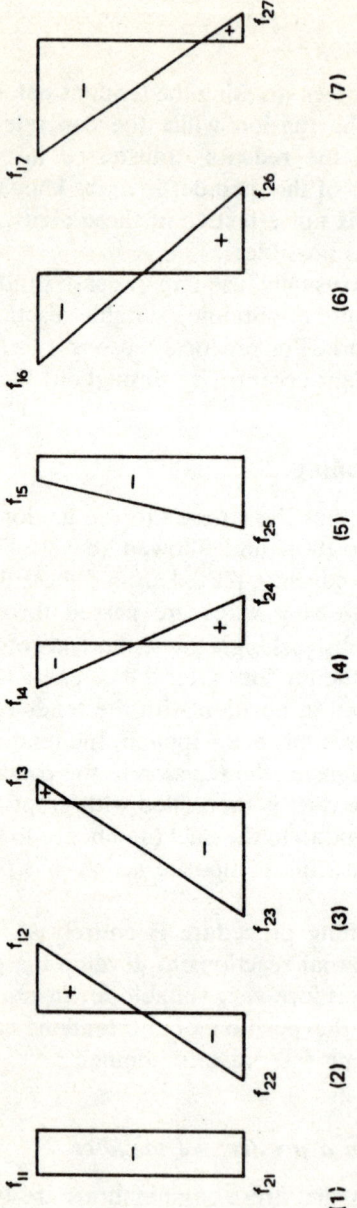

Fig. 5.1 Development of stress in a prestressed member.

developed by the different loads are obvious in this case, the later manipulation of equations to produce limiting conditions of stress means that adherence to a rigid sign convention is an absolute necessity in order to avoid confusion. The sign convention used throughout this chapter is as follows:

1. Tensile stresses and forces are positive.
2. A sagging moment is a positive bending moment.
3. The neutral axis of the section is taken as the y coordinate origin with y being positive below the neutral axis. This means that tendons placed below the NA are at a positive eccentricity, and that the section modulus is positive for elements below the NA and negative for elements above the NA.

In addition to the rigid sign convention, it is necessary to identify the parts of the section that are under discussion. In this context, Face 1 refers to the extreme fibre of the section for a negative y, and Face 2 refers to the extreme fibre of the section for a positive y.

The stresses shown in Fig. 5.1 are:

(1) Uniform compressive stress due to the prestressing force P.

$$f_{11} = f_{21} = \frac{P}{A} \qquad\qquad (5.1)$$

(2) Bending stress due to the eccentricity of the prestressing force P.

$$f_{12} = \frac{Pe}{Z_1} \quad \text{and} \quad f_{22} = \frac{Pe}{Z_2} \qquad (5.2)$$

(3) Combination of (1) and (2). This is the stress-distribution due to the prestressing condition only, but is not a critical condition (see (5) below).

(4) Bending stress due to the self-weight. This occurs at the time of transfer of the prestress into the section, because the action of the prestress moment is to lift the centre of the beam (if a hogging prestress moment) or the ends of the beam (if a sagging moment).

$$f_{14} = \frac{M_D}{Z_1} \quad f_{24} = \frac{M_D}{Z_2} \qquad (5.3)$$

(5) Combination of (3) and (4). This is the overall stress-distribution at transfer and is a steady-state condition requiring consideration in design.

(6) Bending stress due to the live load. A sagging moment is shown so that tensile stresses are produced at Face 2, but the effect of an applied hogging moment is exactly the same except that the stresses are reversed in sign.

$$f_{16} = \frac{M_{LS}}{Z_1} \qquad f_{26} = \frac{M_{LS}}{Z_2} \qquad (5.4)$$

(7) Combination of (5) and (6). This is the situation existing when the beam is under load and, like (5), is a steady-state condition that requires consideration in design.

5.1.4 Loss of prestress

The value of the prestressing force that is required is obtained by considering the steady-state stress conditions in the section and limiting the stresses to certain allowable levels. Over a period of time, however, losses occur in the prestressing force, so that evaluation of the prestress to be applied must consider the potential losses as well as the stress levels in the section.

The losses that occur are due to a number of factors:

1. *Creep in the prestressing steel.* This is known as relaxation, and increases considerably as the tendon stresses approach the ultimate. In order to make allowance for relaxation, some designers assume an arbitrary prestress loss of about 70 N/mm², but BS 8110 recommends a sliding scale which depends on the initial force in the tendon as a proportion of the tendon failure stress. However, if the initial force is less than $0.3f_{pu}$, no allowance is required.
2. *Elastic deformation of the concrete.* In pretensioned members the reduction of the prestressing force due to the elastic deformation of the concrete may be calculated using the modular ratio method, since the strain in the tendons is equal to the strain of elastic deformation of the concrete. Hence, if the concrete is under a stress of f_c, the strain caused by that stress is f_c/E_c. This is equal to the strain change in the tendons, so that the stress loss is $(f_c/E_c)E_s = \alpha_e f_c$.

   In post-tensioned members the problem is not usually so marked and may frequently be allowed for in the sequence of tensioning, i.e. the second and subsequent tendons may be slightly over-tensioned. Alternatively, the loss due to the elastic deformation of the concrete may be entirely eliminated by a slight retensioning.
3. *Shrinkage of concrete.* This is also more important in pretensioned systems, since the transfer of the prestress into the section usually takes place when the concrete is rather younger than in the case of post-tensioned sections. A suggested value of shrinkage strain is given in BS 8110 as 300×10^{-6} for indoor exposure and 100×10^{-6} for outdoor exposure of pretensioned systems, these strains corresponding to prestress losses of approximately 60 N/mm² and 20 N/mm² respectively.
4. *Creep of concrete.* This varies according to the age of the concrete: the

greater the age, the less the creep. The creep strain may be taken as being proportional to the stress in the concrete adjacent to the steel. The actual prestress load depends on the method of tensioning and the exposure conditions, and is summarised in BS 8110.

5. *Anchorage losses.* These occur in post-tensioned members when the transmission of the prestressing force from the jacking system to the anchorages causes 'draw in' of the wedges, or movement of the tendons within the wedges. As it is not a strain, the movement is more important in short members.

6. *Friction.* This also only occurs in post-tensioned members, being the friction between the tendons and the ducts during the prestressing process. The loss in straight tendons is small, but in curved tendons it becomes more important. Lubricants may be used to reduce the frictional effects.

In general, losses are greater in pretensioned systems than in post-tensioned sections, and are typically in the order of $20-25$ per cent of the initial prestressing force for post-tensioned systems, and $30-35$ per cent for pretensioned systems.

In calculating the stress-distributions throughout the section, it is convenient to use the concept of an effective prestress force. This is related to the actual prestressing force by the various losses that may occur, so that the actual prestressing force to be applied is the effective force plus losses. A convenient method of allowing for this is to say that (effective prestress force) = (actual force) × (some constant which is known as the loss ratio). Hence,

$$P_e = P\alpha \qquad (5.5)$$

where α has typical values of between 0·65 and 0·8.

5.2 Design principles

Until the introduction of the limit state approach to design, it was usual to design prestressed members to remain uncracked at all loads, but advances in knowledge and experience have now allowed prestressed members to be classified according to the stresses that may be developed in them under service conditions:

Class 1 No tensile stress exists under the service or working load.

Class 2 The existence of tensile stress is acceptable provided that no visible cracking occurs.

Class 3 Visible cracking is acceptable provided that the crack width does not exceed 0·1 mm for members exposed to particularly aggressive environments, and 0·2 mm for all other members.

There is no hard and fast rule as to the choice of the class of structure that should be used, and in reaching a decision various factors that are specific to the particular design should be considered. In general, a Class 2 structure is suitable for most applications, as it gives the best balance between technical performance and constructional cost, but the use of a Class 1 structure may be specified where a more severe limitation is placed on the deflection, or where some other feature makes a no-crack design particularly essential.

Class 3 structures are frequently referred to as 'partially prestressed' (a term that may also be applied to structures of Class 2), being an intermediate form between the fully prestressed Class 1 condition, in which no tensile cracking occurs, and the simply reinforced section which exhibits extensive cracking in the tensile zone. Class 3 structures are typically used in situations where limited headroom or excessive deflection prohibit the use of a reinforced concrete member.

The design of a prestressed beam involves consideration of both the ultimate and the serviceability limit states. Calculation of the ultimate limit state is, of course, necessary to check the performance of the section against collapse, and therefore takes into account the effects of ultimate loads and bending moments. The serviceability limit state is concerned with the performance of the section under working loads, which involves consideration of any cracking and is therefore affected by the classification of the section.

In general, the limiting condition for Class 1 structures is that of serviceability, and if the section is designed to satisfy this condition it usually also complies with the limit state of collapse without further modification. Class 3 structures are more likely to have the collapse condition as the critical limit state, however, so that, although the initial procedure of designing a Class 1 structure would be:

(a) satisfy the serviceability limit state,
(b) check the limit state of collapse,

for structures of Class 3 the sensible procedure is to reverse the order of the calculation and to check the ultimate condition first. Class 2 structures are best treated in the same sequence as for Class 1, although Class 2 structures satisfying the serviceability limit state often fail to satisfy the ultimate limit state and, like structures of Class 3, require additional reinforcement. This is provided in the form of unpretensioned steel, and is more likely to be required with Class 2 or 3 structures because the acceptance of tensile stresses in the concrete means that a lower section modulus is required, and hence a smaller section. The reduction in section size may be sufficient to prevent the ultimate moment from being carried and, if additional reinforcement is required, the area of unprestressed steel is converted to an equivalent area of prestressed steel so that the total area is the sum of this equivalent area plus the prestressed steel area. The moment of resistance is then determined for the section, taking into account the total steel area.

Once these limit states have been satisfied, the details of prestressing requirements may be determined. The effective prestressing force and the area of tendons required will already have been included in the checking of the serviceability and the collapse limit states, but details of the initial prestress (i.e. consideration of the prestress losses), the design of the tendon anchorages and the transmission length remain. Finally the limit states of shear and deflection must be considered and checked before the design can be considered complete.

5.3   Serviceability limit state

This condition is calculated for the working loads, and is related to the allowable tensile and compressive stresses that may be developed in the concrete and which are shown in Tables 5.1 and 5.2.

Table 5.1   Flexural tensile stresses for Class 2 members: serviceability limit state: cracking (N/mm²)

| | Allowable stress for concrete grade | | | |
	30	40	50	60
Pretensioned members	—	2·9	3·2	3·5
Post-tensioned members	2·1	2·3	2·6	2·8

(Reproduced courtesy of BSI. Source: BS 8110 Part 1, Table 4.1)

Table 5.2   Compressive stresses in concrete for serviceability limit states

Nature of loading	*Allowable compressive stresses*
Design load in bending	$0·33f_{cu}$ In continuous beams and other statically indeterminate structures this may be increased to $0·4f_{cu}$ within the range of support moments
Design load in direct compression	$0·25f_{cu}$

The assumption is that, at the allowable tensile stress, Class 1 and Class 2 structures are uncracked, and so are behaving elastically; design based on linear elastic theory is therefore permissible. Class 3 structures are cracked at this limit state, and should ideally be designed using an elastic theory of cracked sections. No workable theory exists for this situation at present, however, so that the design is based on hypothetical tensile stresses and a theory that is applicable to uncracked sections. The hypothetical stresses are given in Table 5.3. Research has shown that, if these stresses are used with a theory that is applicable to uncracked sections, the limiting crack widths are not exceeded.

Table 5.3   Hypothetical flexural tensile stresses for Class 3 members (N/mm^2)

Group	Limiting crack width (mm)	Stress for concrete grade 30	40	50 and over
A. Pretensioned tendons	0·1	—	4·1	4·8
	0·2	—	5·0	5·8
B. Grouted post-tensioned tendons	0·1	3·2	4·1	4·8
	0·2	3·8	5·0	5·8
C. Pretensioned tendons distributed in the tensile zone and positioned close to the tension faces of the concrete	0·1	—	5·3	6·3
	0·2	—	6·3	7·3

(Reproduced courtesy of BSI. Source: BS 8110 Part 1, Table 4.2)

From Fig. 5.1, the stresses at working load are

$$f_1 = \frac{P_e}{A} + \frac{P_e e}{Z_1} + \frac{M_D}{Z_1} + \frac{M_L}{Z_1} \tag{5.6a}$$

$$f_2 = \frac{P_e}{A} + \frac{P_e e}{Z_2} + \frac{M_D}{Z_2} + \frac{M_L}{Z_2} \tag{5.6b}$$

where M_L is the live load moment and may be either positive (sagging) or negative (hogging). The sagging and hogging live load moments may equally well be considered as the maximum and minimum live load moments respectively, in which case further examination of Fig. 5.1 shows that, for Face 1, the maximum compressive stress occurs under the maximum live load moment, and the maximum tensile stress occurs under the minimum live load moment (assuming that the prestress force and eccentricity remain constant). The converse applies to the stresses in Face 2. The final stress levels therefore depend on both the magnitude and the sign of the live load moment. For these stresses not to exceed the allowable maxima, for Face 1

$$f_{cc_{adm}} \leqslant f_1 = \frac{P_e}{A} + \frac{P_e e}{Z_1} + \frac{M_D}{Z_1} + \frac{M_{L_{max}}}{Z_1} \tag{5.7a}$$

$$f_{ct_{adm}} \geqslant f_1 = \frac{P_e}{A} + \frac{P_e e}{Z_1} + \frac{M_D}{Z_1} + \frac{M_{L_{min}}}{Z_1} \qquad (5.7b)$$

and, for Face 2,

$$f_{cc_{adm}} \leqslant f_2 = \frac{P_e}{A} + \frac{P_e e}{Z_2} + \frac{M_D}{Z_2} + \frac{M_{L_{min}}}{Z_2} \qquad (5.7c)$$

$$f_{ct_{adm}} \geqslant f_2 = \frac{P_e}{A} + \frac{P_e e}{Z_2} + \frac{M_D}{Z_2} + \frac{M_{L_{max}}}{Z_2} \qquad (5.7d)$$

The maximum allowable range of stress in each face is $(f_{ct_{adm}} - f_{cc_{adm}})$ which, from eqs. (5.7a) and (5.7b), gives

$$Z_1 \leqslant \frac{M_{L_{min}} - M_{L_{max}}}{f_{ct_{adm}} - f_{cc_{adm}}} \qquad (5.8a)$$

and, from eqs. (5.7c) and (5.7d),

$$Z_2 \geqslant \frac{M_{L_{max}} + M_{L_{min}}}{f_{ct_{adm}} - f_{cc_{adm}}} \qquad (5.8b)$$

Equations (5.8a) and (5.8b) may therefore be used to give the minimum allowable values of Z_1 and Z_2, so permitting a trial section to be determined. Having chosen a suitable section, the position of the tendons and the magnitude of the prestressing force must be found. Unless the section is symmetric about the neutral axis there is no unique solution of these factors, which are governed by the stress levels given by eqs. (5.7). The prestress moment may be obtained from a high prestress force at a low eccentricity, or from a low prestress force at a high eccentricity. The effects of these alternatives may be seen in Fig. 5.2, which shows that increasing the prestress force increases the final compressive stresses, while reducing the final tensile stresses (assuming that the prestressing moment and all other moments remain constant).

The most economical design is that which uses the minimum prestress at the maximum eccentricity, since the tendon area is reduced. However, it is not always possible to use this combination, as the required eccentricity may be too large for the section, giving insufficient cover to the tendons.

Figure 5.3 compares the effect on the final stress levels of minimum (hogging) and maximum (sagging) live load moments. The effect of increasing the moment is to make the final stress in the top surface of the beam (Face 1) more compressive, and in the bottom surface (Face 2) more tensile, i.e. the total stress line rotates in an anticlockwise direction. The effect of decreasing the moment is the opposite, the total stress line rotating clockwise. Since the effect of decreasing the prestressing force is to reduce the compressive stresses and to increase the tensile stresses in

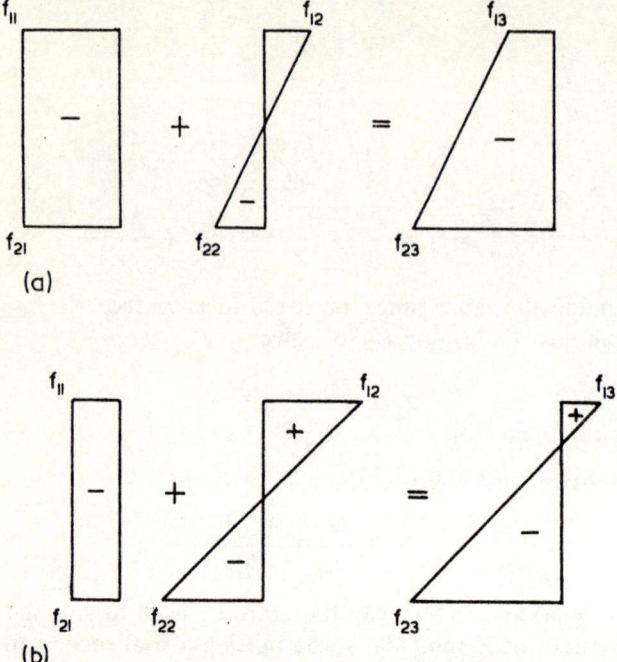

Fig. 5.2 Effect of alternative prestressing configurations on stresses within the section ($Z_1 = -Z_2$). (a) High prestress force, low eccentricity. (b) Low prestress force, high eccentricity.

the section, for a section in which Z_1 is numerically greater than Z_2 the condition governing the minimum prestress force is tension in the bottom face under the maximum live load moment, and either tension in the top face or compression in the bottom face under the minimum live load moment (Fig. 5.3).

The situation in which compression in the top face becomes a limiting condition of minimum prestress can arise only if Z_2 is numerically greater than Z_1, which is unusual in practice. In fact, the limiting condition of compressive stress in either face is not a practical restriction for minimum prestress force, as the total range of stress in either face is governed by the value of the section modulus, the minimum value of which is obtained from eqs. (5.8). In practice, therefore, the conditions governing the minimum prestress force are tension in the bottom face under the maximum moment, and tension in the top face under the minimum moment. Equations (5.7d) and (5.7b) cover this situation, giving

$$P_{e_{\min}} = \frac{A}{Z} (f_{ct_{adm}} \overline{Z} - \overline{M}) \qquad (5.9)$$

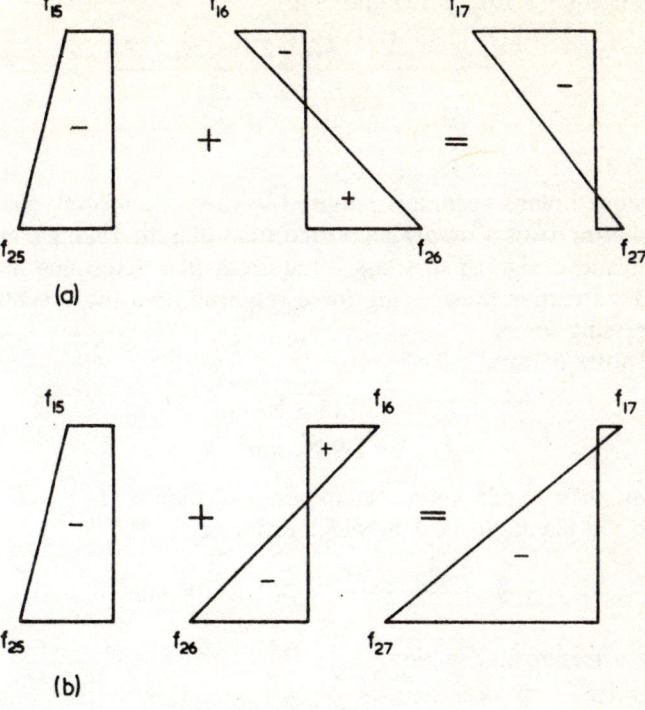

Fig. 5.3 Effect of maximum and minimum live moments on the stresses within the section $(Z_1 > -Z_2)$. (a) Maximum (sagging) live load moment. (b) Minimum (hogging) live load moment.

where $\overline{M}$ is the range of live load moments and $\overline{Z}$ is the numerical sum of Z_1 and Z_2.

Conversely, the maximum prestressing force produces the limiting compressive stress in Face 1 under the maximum live load moment, and in Face 2 under the minimum live load moment. These conditions are covered by eqs. (5.7a) and (5.7c), from which

$$P_{e_{max}} = \frac{A}{\overline{Z}} (f_{cc_{adm}} \overline{Z} - \overline{M}) \qquad (5.10)$$

The eccentricities corresponding to these prestress forces are obtained by substituting the values of P_e given from eqs. (5.9) and (5.10) into the original equations. From eqs. (5.9), (5.7d) and (5.7b),

$$e_{max} = \frac{-M_D \overline{Z} - M_{L_{min}} Z_2 + M_{L_{max}} Z_1}{A (f_{ct_{adm}} \overline{Z} - \overline{M})} \qquad (5.11)$$

while, from eqs. (5.10), (5.7a) and 5.7c),

$$e_{min} = \frac{-M_D \overline{Z} + M_{L_{min}} Z_1 - M_{L_{max}} Z_2}{A(f_{cc_{adm}} \overline{Z} + \overline{M})} \qquad (5.12)$$

Example 5.1

A pretensioned plank section is required to carry a uniformly distributed load of 5 kN/m^2 over a simply supported span of 3 m. Taking concrete of Grade 40, and designing to Class 2 requirements, determine a suitable section, the effective prestressing force required, and the eccentricity of the prestressing force.

From Tables 5.1 and 5.2,

$$f_{cc_{adm}} = -13 \cdot 3 \text{ N/mm}^2$$
$$f_{ct_{adm}} = 2 \cdot 9 \text{ N/mm}^2$$

$M_{L_{max}} = \omega L^2/8 = 5 \cdot 625$ kN m per m width of plank; $M_{L_{min}} = 0$.

Assume the planks to be 1 m wide. From eq. (5.8b),

$$Z_2 \geqslant \frac{5 \cdot 625 \times 10^6}{16 \cdot 2} = 347 \times 10^3 \text{ mm}^3$$

Assuming a rectangular section,

$$Z_1 = -Z_2$$

Therefore

$$Z_1 \leqslant -347 \times 10^3 \text{ mm}^3$$

From these values of Z, the minimum allowable depth is approximately 46 mm. This is likely to prove too small for practical reasons of deflection and tendon spacing, so that the depth might be increased to (say) 60 mm, whence $Z_2 = -Z_1 = 6 \times 10^5$ mm^3, and $A = 60 \times 10^3$ mm^2. From eqs. (5.9) and (5.10),

$$P_{e_{min}} = -107 \text{ kN}$$

(The negative sign shows that the prestress is compressive. This follows the sign convention of Section 5.1.3.)

$$P_{e_{max}} = -516 \cdot 75 \text{ kN}$$

Calculation of the maximum and minimum tendon eccentricities requires the dead load bending moment. Assuming density of concrete to be 2400 kg/m^3,

$$\text{dead load} = 0 \cdot 06 \times 1 \times 2 \cdot 4 \times 9 \cdot 81 \text{ kN/m over full span}$$
$$= 1 \cdot 41 \text{ kN/m}$$

Therefore maximum dead load moment

$$M_{D_{max}} = \frac{1 \cdot 41 \times 9}{8} = 1 \cdot 59 \text{ kN m}$$

From eq. (5.11),

$$e_{max} = 41 \text{ mm}$$

and from eq. (5.12),

$$e_{min} = 8 \cdot 52 \text{ mm}$$

After allowing for cover and the size of the prestressing wires, the maximum eccentricity that can be used is about 13 mm. Therefore use 516·75 kN prestressing force at eccentricity of 8·5 mm on a section 1 m wide and 60 mm deep.

N.B. It is possible to reduce the prestressing force in the section in order to take advantage of the extra eccentricity that is available (e may be increased to about 13 mm). The calculation associated with such a procedure is discussed in the next section.

5.4   The permissible tendon zone

The structure is designed to satisfy the limiting stress conditions under maximum dead and live load moments, so that any reduction in these moments will mean that the section has excess capacity at that point, and that the stresses within the section will be below their critical or limiting values. In practical terms this means that the prestress moment may be varied without exceeding the critical stresses, the amount of variation being dependent upon the maximum moments occurring within the span and the moments at the section under consideration. (This assumes that the section has constant cross-sectional shape and dimensions.) Assuming that the prestressing force remains constant, the variation in prestress moment may be obtained by varying the positions of the tendons. The limiting allowable values of the stresses must not be exceeded, however; eqs. (5.7) may be used to ensure this, but it is more convenient to rewrite these equations to give values of the tendon eccentricity that correspond to the limiting stresses being reached in the section, so that the limiting eccentricities are given by eqs. (5.13):

$$e \geqslant \frac{f_{cc_{adm}} Z_1}{P_e} - \frac{Z_1}{A} - \frac{M_D}{P_e} - \frac{M_{L_{max}}}{P_e} \qquad (5.13a)$$

$$e \leqslant \frac{f_{ct_{adm}} Z_1}{P_e} - \frac{Z_1}{A} - \frac{M_D}{P_e} - \frac{M_{L_{min}}}{P_e} \qquad (5.13b)$$

$$e \geqslant \frac{f_{ct_{adm}} Z_2}{P_e} - \frac{Z_2}{A} - \frac{M_D}{P_e} - \frac{M_{L_{max}}}{P_e} \qquad (5.13c)$$

$$e \leqslant \frac{f_{cc_{adm}} Z_2}{P_e} - \frac{Z_2}{A} - \frac{M_D}{P_e} - \frac{M_{L_{min}}}{P_e} \qquad (5.13d)$$

These equations define a zone within the section in which the tendons may be placed anywhere without exceeding the allowable stresses. The zone is known as the permissible tendon zone, and it will be found that the equations governing the critical maximum and minimum tendon eccentricity at any cross-section are the limiting equations for all points along the span. This shown in example 5.2.

Example 5.2
Calculate the permissible tendon zone for the section of Example 5.1 for an effective prestressing force of 400 kN.

At mid-span, $M_D = 1 \cdot 59$ kN m, $M_{L_{max}} = 5 \cdot 625$ kN m, $M_{L_{min}} = 0$. From eqs. (5.13),

(a) $e \geqslant \dfrac{-13 \cdot 3 \times -6 \times 10^5}{-400 \times 10^3} - \dfrac{-6 \times 10^5}{60 \times 10^3} - \dfrac{1 \cdot 59 \times 10^6}{-400 \times 10^3} - \dfrac{5 \cdot 625 \times 10^6}{-400 \times 10^3}$

giving $e \geqslant 8 \cdot 1$ mm.

(b) $\qquad e \leqslant \dfrac{2 \cdot 9 \times -6 \times 10^5}{-400 \times 10^3} - \dfrac{-6 \times 10^5}{60 \times 10^3} - \dfrac{1 \cdot 59 \times 10^6}{-400 \times 10^3}$

giving $e \leqslant 18 \cdot 3$ mm.

(c) $\quad e \geqslant \dfrac{2 \cdot 9 \times 6 \times 10^5}{-400 \times 10^3} - \dfrac{6 \times 10^5}{60 \times 10^3} - \dfrac{1 \cdot 59 \times 10^6}{-400 \times 10^3} - \dfrac{5 \cdot 625 \times 10^6}{-400 \times 10^3}$

giving $e \geqslant 3 \cdot 69$ mm.

(d) $\qquad e \leqslant \dfrac{-13 \cdot 3 \times 6 \times 10^5}{-400 \times 10^3} - \dfrac{6 \times 10^5}{60 \times 10^3} - \dfrac{1 \cdot 59 \times 10^6}{-400 \times 10^3}$

giving $e \leqslant 13 \cdot 9$ mm.

From eqs. (5.13a) and (5.13d), 8·1 mm $< e <$ 13·9 mm at midspan. At quarter span, $M_D = 1 \cdot 19$ kN m, $M_{L_{max}} = 4 \cdot 22$ kN m, $M_{L_{min}} = 0$.
From eqs. (5.13),

(a) $e \geqslant 3 \cdot 6$ mm
(b) $e \leqslant 17 \cdot 3$ mm
(c) $e \geqslant -0 \cdot 8$ mm
(d) $e \leqslant 12 \cdot 9$ mm

Therefore, from eqs. (a) and (d), 3·6 mm $< e <$ 12·9 mm at quarter span.

At the support, $M_D = M_{L_{max}} = M_{L_{min}} = 0$. Once again, using eqs. (5.13),

(a) $e \geqslant -9.95$ mm
(b) $e \leqslant 14.35$ mm
(c) $e \geqslant -14.35$ mm
(d) $e \leqslant 9.95$ mm

Therefore, from (a) and (d), -9.95 mm $< e < 9.95$ mm at supports.

Although the tendon eccentricities have been calculated for each of the possible limiting stress conditions, it will be noted that at each section considered the limiting condition is given by the same equations—in this case eqs. (5.13a) and (5.13d). The permissible tendon zone is shown in Fig. 5.4. Its position within the section means that frequently in post-tensioned structures the actual tendon profile is inclined to the horizontal.

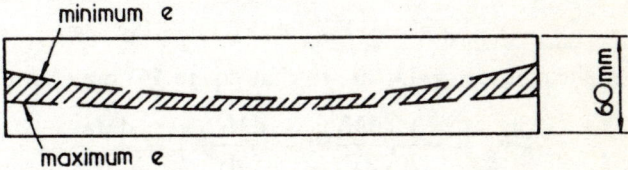

Fig. 5.4 Permissible tendon zone.

This inclination produces a vertical component of the tendon force which is proportional to the sine of the angle of inclination, so that

$$V_p = -P_e \sin \theta \qquad (5.14)$$

where V_p is the vertical component of the tendon force. For small angles,

$$\sin \theta \simeq \frac{de}{dx}$$

so that

$$V_p = -P_e \frac{de}{dx} \qquad (5.15)$$

This shear force acts in opposition to the applied shear forces on the section (hence the negative sign), so that the width of the permissible tendon zone may be used to vary the inclination of the tendons according to the magnitude of the shear forces acting at a particular section, and so reduce the effective shear force at the section concerned. In general,

$$V_c = V + V_p$$

where V_c is the resultant shear force on the concrete section, and V is the shear force applied to the section:

$$V = V_{\text{live loads}} + V_{\text{dead loads}} \equiv V_{\text{LL}} + V_{\text{D}}$$

It will frequently be found that the load system producing the critical bending moments also produces the greatest shear forces. In such cases the extremes of V_c are given by

$$V_{L_{\text{max}}} + V_{\text{D}} + V_{\text{p}}$$
$$V_{L_{\text{min}}} + V_{\text{D}} + V_{\text{p}}$$

The optimum tendon profile (for the shear condition) is that which reduces the value of V_c to a minimum, in which case

$$V_{L_{\text{max}}} + V_{\text{D}} + V_{\text{p}} = -(V_{L_{\text{min}}} + V_{\text{D}} + V_{\text{p}})$$

giving

$$V_{\text{p}} = -\tfrac{1}{2}(V_{L_{\text{max}}} + V_{L_{\text{min}}}) - V_{\text{D}} \tag{5.16}$$

However, shear force $= dM/dx$, so that eq. (5.16) may be rewritten

$$V_{\text{p}} = -\frac{1}{2}\left(\frac{dM_{L_{\text{max}}}}{dx} + \frac{dM_{L_{\text{min}}}}{dx}\right) - \frac{dM_{\text{D}}}{dx} \tag{5.17}$$

and, from eqs. (5.15) and (5.17),

$$-P_e\frac{de}{dx} = -\frac{1}{2}\left(\frac{dM_{L_{\text{max}}}}{dx} + \frac{dM_{L_{\text{min}}}}{dx}\right) - \frac{dM_{\text{D}}}{dx} \tag{5.18}$$

Integrating both sides, and rearranging, gives

$$e = \frac{1}{P_e}\left(\frac{M_{L_{\text{max}}}}{2} + \frac{M_{L_{\text{min}}}}{2} + M_{\text{D}}\right) + K \tag{5.19}$$

where K is a constant of integration, the value of which should be chosen to ensure that the tendon profile for optimum shear lies within the permissible tendon zone. The tendon must always lie within the permissible zone, so that if the tendon profile for optimum shear does not coincide with the permissible zone, the actual position of the tendons must conform to the latter condition and not to the former.

5.5   Limit state of collapse

The method of obtaining the ultimate moment of a prestressed section is very similar to the procedure used for reinforced concrete sections, except that the stress in the concrete may be determined from consideration of either the rectangular parabolic stress distribution derived from Fig. 3.4, or the more idealised simple rectangular stress block as used in the earlier treatment of reinforced concrete beams. The design equations for the collapse of a prestressed section may therefore be constructed in the same way as those for a reinforced concrete beam, and are identical in form with those previously obtained:

$$M_u = 0.4f_{cu}bx(d - d_n) \qquad (5.20)$$

$$M_u = f_{pb}A_{ps}(d - d_n) \qquad (5.21)$$

where f_{pb} is the stress in the tendons at failure, and d_n is the depth to the centroid of the compression zone. For the rectangular stress block, $d_n = 0.45x$.

Determination of the neutral axis position is not as simple as for the reinforced beam, since the steel has some prestress strain that must be considered in obtaining the final strain profile. However, the basic assumption of plane sections remaining plane is maintained, so that a linear strain diagram may be used and, assuming that the strain profile for a simple rectangular prestressed section under a flexural load is as shown in Fig. 5.5, the neutral axis position may be found.

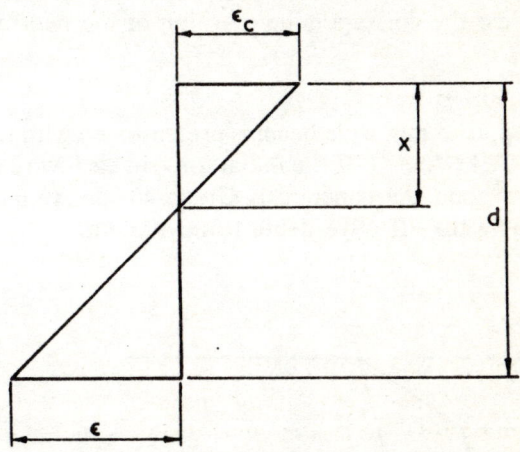

Fig. 5.5   Assumed strain profile for a rectangular section under flexure.

The final strain in the steel is made up of the initial strain due to the prestressing force (after due allowance for the various losses that may occur), plus the additional strain due to the load application:

$$\epsilon_p + \epsilon = \epsilon_s$$

From Fig. 5.5,

$$\epsilon = \frac{\text{strain in the concrete }(d - x)}{x}$$

so that, assuming the limiting concrete strain to be 0·0035,

$$\epsilon = \frac{0 \cdot 0035(d - x)}{x} \tag{5.22}$$

The stress in the steel may be evaluated from knowledge of the strain and the stress-strain relationship. Although this is a potentially simple exercise, it is complicated by the non-linearity of the stress-strain curve, which is assumed to be as shown in Fig. 5.6. It is therefore not practicable to write a general expression for the stress in the steel in terms of the strain, so the steel stress must be evaluated using the stress-strain curve. As the steel area is known, this then allows the force in the steel to be calculated, and for equilibrium the force in the steel must be balanced by the force in the concrete. The depth to the neutral axis may then be evaluated, and the ultimate moment determined.

In practice, it is convenient to assume a neutral axis position and to compare the compressive and tensile forces that correspond to this NA. This involves determining the tensile force by calculating the value of strain in the steel corresponding to the neutral axis depth; an alternative is to assume the value of the steel strain and then to use this assumed value to calculate the corresponding position of the neutral axis.

Example 5.3
A 125 mm deep, 63·5 mm wide beam is pretensioned with three 5 mm dia wires ($f_{pu} \simeq 1·57$ kN/mm^2). If the initial force in each wire after all losses is 0·96 kN/mm^2, and the concrete is Grade 40, determine the ultimate moment. Assume the effective depth to be 100 mm.

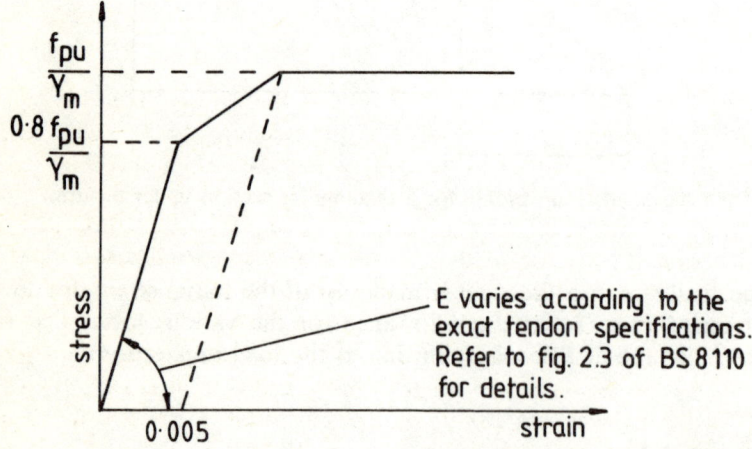

Fig. 5.6 Typical stress-strain curve for prestressing tendons.

The section is as shown in Fig. 5.7, and the stress-strain relationship of the prestressing tendons may be taken as that shown in Fig. 5.6. The initial strain in the prestressing steel

$$\epsilon_p = \frac{0.96}{200} = 0.0048$$

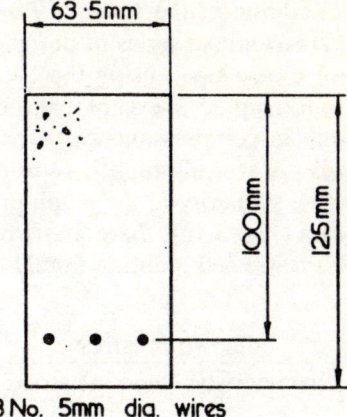

3 No. 5mm dia. wires

Fig. 5.7

Assuming a neutral axis depth of 50 mm, from eq. (5.22) steel strain due to the load application

$$\epsilon = \frac{0.0035(100 - 50)}{50}$$

$$= 0.0035$$

Therefore the *total* steel strain = 0·0035 + 0·0048 = 0·0083.

From the stress-strain curve (Fig. 5.6) and assuming that E = 200 kN/mm², stress in the steel = 1·214 kN/mm², therefore total force in steel = 1·214 × 58·8 = 71·4 kN.

Using the rectangular stress block for the concrete in compression, the total compressive force = 0·45 × 40 × 0·9 × 50 × 63·5 = 50·8 kN. Hence, as $F_c < F_t$, the assumed depth of the neutral axis is too low, and the actual value of x is greater than that tried.

Repeating the above calculation for a neutral axis depth of 66 mm gives a total compressive force F_c of 67·06 kN, and a total tensile force F_t of 67·03 kN. The lever arm for this neutral axis position is 100 − 30 = 70 mm, so that the ultimate moment capacity of the section is

$$M_u = 67.03 \times 70 = 4.69 \text{ kN m}$$

Evaluation of the ultimate moment by this method is rather tedious for design applications, since it involves trial-and-error determination of the neutral axis position. Chapter 3 showed that graphical solutions could be produced for rectangular reinforced concrete beams that allowed the design procedure to be considerably shortened, and similar graphical solutions may also be developed for the prestressed rectangular beam. The graphs are best presented for various characteristic concrete and steel strengths, and the 1972 edition of the *Code of Practice for the structural use of concrete* (CP 110) included a series of design charts for prestressed sections. The charts were developed using the basic procedure outlined above, and illustrated in Example 5.3, except that the rectangular parabolic stress block for concrete in compression was used in preference to the simple rectangular block. A typical graph is reproduced in Fig. 5.8, and although this bears a close similarity to the graph produced for the simple reinforced concrete beam (Fig. 3.12), there are two principal differences.

One is that, on each prestressed graph, a family of curves is drawn for various ratios of

$$\frac{\text{effective prestress}}{\text{characteristic (tendon) stress}}$$

This is to take into account the losses that occur in the prestressing procedure, and which may vary from section to section according to the factors discussed earlier in this chapter. The other main difference concerns the discontinuity of the curve which appears in the prestress graphs. This is due to the non-linearity of the assumed stress-strain curve for the prestressing tendons; although the stress-strain relationship for reinforcing steels exhibits a similar discontinuity, this does not appear in the design graphs for reinforced sections because calculation of the ultimate moment of such sections assumes that the reinforcement strain is sufficiently high to correspond with the upper linear part of the curve. This assumption cannot be made for prestressed sections, however, so that the discontinuity appears in the design graph.

The graphs are very simple to use. As an example of their use, the section of Example 5.3 is reconsidered here using the graphical approach.

Example 5.4
From the section specification,

$$\frac{100A_{ps}}{bd} = \frac{100 \times 58 \cdot 8}{63 \cdot 5 \times 100} = 0 \cdot 926$$

and

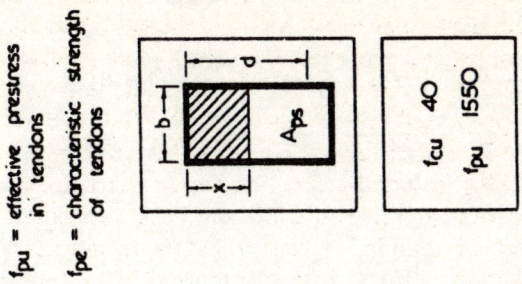

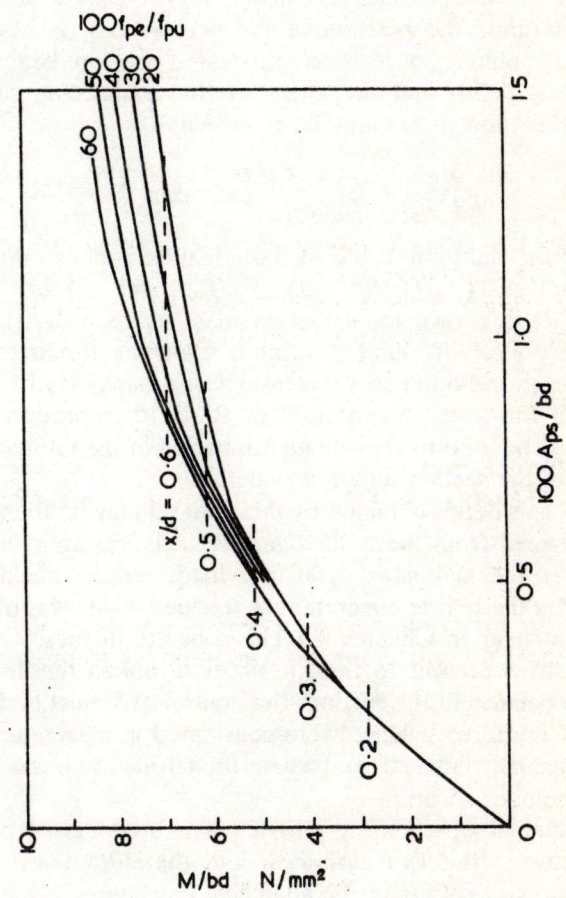

Fig. 5.8 Design graph for prestressed sections (from CP 110).

$$\frac{100f_{pe}}{f_{pu}} = \frac{100 \times 0 \cdot 96}{1 \cdot 57} = 61$$

From Fig. 5.8, for

$$\frac{100A_{ps}}{bd} = 0 \cdot 926 \qquad \text{and} \qquad \frac{100f_{pe}}{f_{pu}} = 61$$

$$\frac{M}{bd^2} = 7 \cdot 3 \text{ N/mm}^2$$

giving

$$M_u = 4 \cdot 63 \text{ kN m}$$

An alternative procedure which eliminates the necessity of analysing the section to determine the position of the neutral axis, is based on the results of a large number of tests on prestressed concrete beams in which the neutral axis depth and the stress in the tendons at failure were determined. Equation (5.21) may be rewritten

$$\frac{M_u}{bd^2 f_{cu}} = \frac{f_{pb}A_{ps}f_{pu}}{f_{cu}bdf_{pu}} \left(1 - \frac{x}{2 \cdot 2d}\right) \tag{5.23}$$

in which case the right-hand side of the equation contains three dimensionless groupings, $(A_{ps}f_{pu})/(f_{cu}bd)$, f_{pb}/f_{pu}, and $x/2 \cdot 2d$. Of these, $(A_{ps}f_{pu})/(f_{cu}bd)$ is known for a section that has been designed on the basis of the serviceability limit state and, using the results of the tests mentioned above, the other two dimensionless groups may be evaluated. The results of the tests, summarised in BS 8110 (reproduced here as Table 5.4), may be used to provide an estimation of the ultimate moment of resistance of the section under consideration.

The neutral axis depth obtained by this method may be found to differ from that obtained from the more rigorous analytical approach. This is because, under the action of a flexural load, tensile cracking of the concrete divides the tensile concrete into tongues. (This was discussed in the section on shear in Chapter 4.) The concrete in these tongues has some tensile stress present so that, in order to obtain tensile and compressive force balance in the section, the neutral axis must be lower than it would be if the force balance were considered at a section where the concrete was actually cracked, and where the tensile force was due to the reinforcement only.

The experimental observations provide a neutral axis depth that is some way between the two extremes, and therefore tends to give a greater neutral axis depth than the analytical procedure, which considers the depth to the head of the tension cracks. This means that the lever arm calculated from the above method is less than that obtained from a

Table 5.4 Conditions at the ultimate limit state for rectangular beams with pretensioned tendons or with post-tensioned tendons having effective bond

$\dfrac{f_{pu}A_{ps}}{f_{cu}bd}$	Design stress in tendons as a proportion of the design strength, $f_{pb}/0.87f_{pu}$			Ratio of depth of neutral axis to that of the centroid of the tendons in the tension zone, x/d		
	$f_{pe}/f_{pu} =$			f_{pe}/f_{pu}		
	0.6	0.5	0.4	0.6	0.5	0.4
0.05	1.0	1.0	1.0	0.11	0.11	0.11
0.10	1.0	1.0	1.0	0.22	0.22	0.22
0.15	0.99	0.97	0.95	0.32	0.32	0.31
0.20	0.92	0.90	0.88	0.40	0.39	0.38
0.25	0.88	0.86	0.84	0.48	0.47	0.46
0.30	0.85	0.83	0.80	0.55	0.54	0.52
0.35	0.83	0.80	0.76	0.63	0.60	0.58
0.40	0.81	0.77	0.72	0.70	0.67	0.62
0.45	0.79	0.74	0.68	0.77	0.72	0.66
0.50	0.77	0.71	0.64	0.83	0.77	0.69

(Reproduced courtesy of BSI. Source: BS 8110 Part 1, Table 4.4)

rigorous analysis, and the ultimate moment may therefore also be expected to be less than that given by the rigorous method. The increased neutral axis depth increases the force in the compression zone, however, and this requires an increased tensile force for equilibrium, so that the actual ultimate moment does not show much variation from that obtained analytically. For Example 5.5 it will be seen that the ultimate moment calculated in this way is in fact very close to that given by the analytical method.

Example 5.5
Determine the ultimate moment capacity of the section detailed in Example 5.3, but using eq. (5.21) and the information given in Table 5.4.

From the section specification, $\dfrac{f_{pe}}{f_{pu}} = 0.61$,

and $\qquad \dfrac{A_{ps}f_{pu}}{f_{cu}bd} = \dfrac{58.8 \times 1.57 \times 10^3}{40 \times 63.5 \times 100} = 0.363$

From Table 5.4, for $(A_{ps}f_{pu})/(f_{cu}bd)$ of this value and $\dfrac{f_{pe}}{f_{pu}} = 0.61$,

$$\frac{f_{pb}}{0.87 f_{pu}} = 0.82 \qquad \text{and} \qquad \frac{x}{d} = 0.65$$

Therefore $f_{pb} = 1.12$ kN/mm^2 and $x = 65$ mm, so that substituting into eq. (5.21) gives

$$M_u = 4.66 \text{ kN m}$$

The above discussion of the ultimate moment capacity of a prestressed section relates to the situation in which the tendon is either pretensioned or post-tensioned and grouted. In the case of a section having a post-tensioned ungrouted tendon, calculation of the ultimate moment is complicated by difficulty in determining the tendon strain. In the grouted or pretensioned tendon the maximum strain occurs at the point of ultimate moment, whereas in the ungrouted state the strain is constant all along the tendon length. Determination of the strain at ultimate load for the ungrouted tendon therefore involves summing the strain at all sections along the length of the member.

For design applications the procedure is made more simple by utilizing the equation presented in BS 8110 for the design tensile stress in the tendons. From this,

$$f_{pb} = f_{pe} + \frac{7000}{l/d}\left(1 - 1.7\frac{f_{pu}A_{ps}}{f_{cu}bd}\right) \tag{5.24}$$

where l is the length of the tendons between the end anchorages. The value of x may be obtained by equating longitudinal forces,

$$0.405 f_{cu} bx = A_{ps}f_{pb}$$

giving

$$x = \frac{2.47 A_{ps}f_{pb}}{f_{cu}b} \tag{5.25}$$

This is represented in BS 8110 by the equation

$$x = 2.47\left[\left(\frac{f_{pu}A_{ps}}{f_{cu}d}\right)\left(\frac{f_{pb}}{f_{pu}}\right)d\right]$$

which is equation (5.25) multiplied by $\dfrac{f_{pu}d}{f_{pu}d}$ and rearranged.

The values of x and f_{pb} can then be substituted into eq. (5.21) to give the ultimate moment capacity of a prestressed section with ungrouted tendons. However, it should be realized that at the critical section, the strain in the tendons is much lower than the strain in the concrete.

This means that the ultimate moment capacity of the beam is not properly developed, and a prestressed beam having ungrouted tendons will fail at a moment that is lower than that for a similar section having grouted tendons. In addition, the ungrouted section will exhibit a larger deflection and wider cracking so that the use of ungrouted tendons is not recommended.

5.6 Stresses at transfer

Although the serviceability limit state is concerned with the stress combinations under working loads, notice must also be taken of the stresses within the section at the time of transfer. This occurs when the prestressing force is actually applied to the section and produces the stress-distributions shown in Fig. 5.1(5), at which time the stresses in the section are given by:

$$f_{15} = \frac{P}{A} + \frac{Pe}{Z_1} + \frac{M_D}{Z_1} \qquad (5.26a)$$

$$f_{25} = \frac{P}{A} + \frac{Pe}{Z_2} + \frac{M_D}{Z_2} \qquad (5.26b)$$

The limit state at transfer is safeguarded by ensuring that stresses f_1 and f_2 lie between the allowable limits of tensile and compressive stress for the structure in question. The allowable compressive stresses are shown in Table 5.5, while the allowable tensile stresses should not exceed 1 N/mm^2 for a Class 1 structure, and $0.45\sqrt{f_{cu}}$ for pretensioned or $0.36\sqrt{f_{cu}}$ for post-tensioned structures of Classes 2 and 3.

Table 5.5 Allowable compressive stresses at transfer

Nature of stress distribution	Allowable compressive stresses
Triangular or near triangular distribution of prestress	$0.5f_{ci}$
Uniform or near uniform distribution of prestress	$0.4f_{ci}$

f_{ci} is the concrete strength at transfer.

It may be seen from Table 5.5 that the proportion of the characteristic concrete stress giving the allowable transfer stress is higher than the proportion for the general serviceability condition. This allows for the reduction in load factor that is acceptable for a temporary load such as occurs during the period of construction, and which will either be removed or modified when the working loads are applied. Further reference to

Table 5.5 also shows that the allowable stresses are given in terms of the concrete strength at transfer, and not the concrete characteristic strength. This is to allow for the age of the concrete at the time of prestressing, which may occur when the concrete is comparatively 'young' and has not yet developed the full potential of the specified characteristic strength. With the exception of Class 1 structures, the allowable tensile stresses are not modified over those for the serviceability condition.

Restriction of the stresses at transfer to allowable levels implies that, as with the serviceability limit state, there is a limiting combination of prestress force and tendon eccentricity. Assuming that the prestress force remains constant at the value previously determined from the serviceability calculations, the permissible eccentricity for satisfaction of the transfer stresses may be obtained from eqs. (5.26). Rewriting these equations gives:

$$e = \frac{f_{15}Z_1}{P} - \frac{Z_1}{A} - \frac{M_D}{P} \qquad (5.27a)$$

$$e = \frac{f_{25}Z_2}{P} - \frac{Z_2}{A} - \frac{M_D}{P} \qquad (5.27b)$$

The limiting values of eccentricity may be determined by setting both f_{15} and f_{25} equal to the allowable tensile and compressive stresses, which gives:

$$e \geqslant \frac{f_{cc_{adt}}Z_1}{P} - \frac{Z_1}{A} - \frac{M_D}{P} \qquad (5.28a)$$

$$e \leqslant \frac{f_{ct_{adt}}Z_1}{P} - \frac{Z_1}{A} - \frac{M_D}{P} \qquad (5.28b)$$

$$e \geqslant \frac{f_{ct_{adt}}Z_2}{P} - \frac{Z_2}{A} - \frac{M_D}{P} \qquad (5.28c)$$

$$e \leqslant \frac{f_{cc_{adt}}Z_2}{P} - \frac{Z_2}{A} - \frac{M_D}{P} \qquad (5.28d)$$

As with the corresponding situation with the serviceability limit state, the values of e may be found for various points along the beam, thus defining a permissible tendon zone for the stresses at transfer. There is no need to calculate all four values of eccentricity for each point considered along the beam, for again it will be found that the equations giving the limiting eccentricity of the tendons at the centre of the span will also be the critical equations at other sections. Notice that, in considering the stress conditions at transfer, the prestressing force used is that applied by the jacks, and makes no allowance for the various losses that may occur within the system.

5.7 Shear in prestressed beams

The shear performance of the section concerns the ultimate limit state condition only, and no calculation is required for the serviceability condition. The effect of a prestressing force on the shear behaviour is twofold. Firstly, the presence of the prestress reduces the flexural cracking and therefore also the formation of flexural shear cracks; secondly, the widths of any flexural cracks that do form are reduced when compared with the reinforced case, so that the aggregate interlock forces are increased. The presence of a prestressing force therefore increases the resistance of the section to shear, and allows thinner web sections to be used.

Evaluation of the shear performance of a prestressed section involves consideration of two conditions, either of which may arise in any one section: (a) the section is uncracked in flexure; (b) the section is cracked in flexure. These two conditions may arise simultaneously in any one section, depending upon the distribution of bending moments along the section. A beam must therefore be checked for both.

5.7.1 Section uncracked in flexure

In this condition, shear cracks occur at the centre of the beam but do not extend to the tension face. Assuming that any element of the beam at the centroid of the section is subjected to a longitudinal compressive stress (due to the prestress force) of f_{cp} and a shear stress v, then, from elastic theory,

$$f_t = -\tfrac{1}{2}f_{cp} + \tfrac{1}{2}\surd(f_{cp}^2 + 4v^2) \qquad (5.29)$$

where f_t is the principal tensile stress at the centroid.

The maximum shear stress resulting from the application of a shear force to a rectangular section is

$$v = 1\cdot5\,\frac{V}{bh}$$

so that putting $V_{co} = V$ and $v_{co} = v$, and substituting into eq. (5.29), gives

$$V_{co} = 0\cdot67bh\,(f_t^2 + f_t f_{cp})^{1/2} \qquad (5.30)$$

Application of a partial safety factor of 0·8 to the value of f_{cp} then makes the equation

$$V_{co} = 0\cdot67bh\,(f_t^2 + 0\cdot8f_t f_{cp})^{1/2} \qquad (5.31)$$

There is no necessity to include a partial factor on the value of f_t, since this is effectively considered by putting an upper limit on f_t of $0\cdot24\,\surd f_{cu}$.

5.7.2 Section cracked in flexure

There is no simple theory at present to provide for ultimate shear performance of sections that are cracked in flexure; BS 8110 uses an equation that is derived from the results of a large number of tests:

$$V_{cr} = \left(1 - \frac{0 \cdot 55 f_{pe}}{f_{pu}}\right) v_c bd + \frac{M_0}{M} V \qquad (5.32)$$

where M_0 = moment producing zero stress in the concrete at depth d.
f_{pe} = effective prestress after all losses ($f_{pe} \not> f_{pu}$ in this equation)
V and M are the shear force and bending moment at the section considered (due to ultimate loads)

Having examined both the cracked and the uncracked conditions, the critical shear on the section is taken as V_c where V_c is numerically equal to the lesser of V_{cr} and V_{co}. If the shear force acting on the section V is less than $0 \cdot 5 V_c$, no shear reinforcement is required. If $0 \cdot 5 V_c < V > (V_c + 0 \cdot 4bd)$, minimum shear reinforcement should be provided such that

$$\frac{A_{sv}}{s_v} = \frac{0 \cdot 4b}{0 \cdot 87 f_{yv}} \qquad (5.33)$$

If $V > (V_c + 0 \cdot 4bd)$, reinforcement should be provided to carry the additional shear so that

$$\frac{A_{sv}}{s_v} = \frac{V - V_c}{0 \cdot 87 f_{yv} d_t} \qquad (5.34)$$

where d_t is the effective depth in shear. For both cracked and uncracked sections there is an upper limit on the maximum design shear stress given by the lesser of $0 \cdot 8(f_{cu})$ or 5 N/mm^2. This is to avoid the possibility of failure by diagonal compression.

The basic design of the prestressed member may therefore be considered to be complete, with only the details of anchorage, end block design, deflection and transmission length to be considered. The design steps may therefore be summarised:

1. From consideration of the serviceability limit state, determine (a) a suitable cross-section, (b) the effective prestressing force and the eccentricity, (c) the permissible tendon zone.
2. After allowing for the various losses that are likely to occur in the prestressing force, choose a suitable tendon size. Note that the maximum initial prestress should not normally exceed 0·7 of the characteristic tendon strength, unless additional consideration is given to safety, etc.
3. Check the ultimate state of flexure.
4. Check the transfer stresses and the tendon profile for transfer conditions.

5. Check the ultimate limit state of shear.
6. Check deflection, design of end blocks, and other detail points.

For a Class 3 structure, Step 3 should be considered before Step 1.

Example 5.6 works through most of this procedure. For simplicity, however, the calculations are restricted to the centre point of the span, which is the critical condition for bending but at which the shear forces are effectively zero.

Example 5.6
As an illustration of the complete design procedure, consider the design of a Class 2 beam that is required to carry a uniformly distributed load of 15 kN/m over a simply supported span of 10 m. Load factors of 1·6 on live loads and 1·4 on dead loads are to be taken, with the concrete assumed to be of Grade 50, and an allowance of 20 per cent made for the prestress losses.

From Tables 5.1 and 5.2, for Grade 50 concrete, $f_{cc_{adm}} = -16\cdot5$ N/mm^2, $f_{ct_{adm}} = 2\cdot55$ N/mm^2. $M_{L_{max}} = \omega L^2/8 = 187\cdot5$ kN m. From eq. (5.8), the minimum section modulus required

$$Z_{min} = \frac{187 \times 10^6}{19\cdot05} = 9\cdot84 \times 10^6 \text{ mm}^3$$

Try the section shown in Fig. 5.9:

$$Z_1 = -12\cdot5 \times 10^6 \text{ mm}^3$$
$$Z_2 = 11\cdot2 \times 10^6 \text{ mm}^3$$
$$A = 90 \times 10^3 \text{ mm}^2$$

The dead load may now be evaluated; at the centre of the span, $M_D = 26\cdot5$ kN m.

From eqs. (5.9) and (5.11),

$$P_{e_{min}} = -482 \text{ kN}$$
$$e_{max} = 260 \text{ mm}$$

The minimum prestress force at the maximum eccentricity may therefore be used, as the depth of the section still provides adequate cover to the tendons. However, in order to allow for some latitude in the construction, use P_e of 500 kN. From eqs. (5.13), considering the midspan position,

(a) $e \geqslant 154$ mm
(b) $e \leqslant 256$ mm
(c) $e \geqslant 246$ mm
(d) $e \leqslant 298$ mm

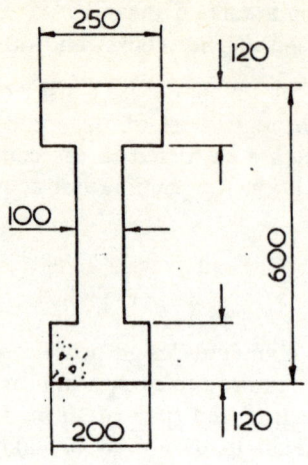

All dimensions in mm

Fig. 5.9

so that, at this cross-section, 246 mm $\leqslant e \leqslant$ 256 mm. Therefore take e to be 250 mm at this section.

The critical conditions of tendon eccentricity are given by eqs. (5.13b) and (5.13c), which are now applied to various points on the span in order to determine the tendon profile. Since this is a purely repetitive calculation, it is not performed here.

The tendon size may now be chosen. Allowing for losses, the initial prestressing force must be 625 kN. Since this should not exceed $0·7f_{pu}A_{ps}$, a minimum characteristic strength of 892 kN is required. One 35 mm dia prestressing bar, having a specified characteristic strength of 950 kN, may thus be used. This will be set at an eccentricity (at midspan) of 250 mm so that, at the point of maximum moment, the effective depth = 550 mm.

The initial prestress (after all losses) = 500/961 = 0·52 kN/mm², so that the initial strain ϵ_p = 0·52/200 = 0·0026.

Following the procedure of Section 5.5, and assuming the stress-strain curve for the steel to be as shown in Fig. 5.6, the neutral axis depth is found to be approximately 200 mm, which gives F_c = 815 kN and F_t = 790 kN. From eq. 5.21, therefore, M_U = 810(550 − 90) = 372 kN m.

The ultimate moment required = 187·5 × 1·6 × 26·5 × 1·4 = 337 kN m so that the section satisfies the ultimate limit state of flexure.

Transfer stresses. Assuming that the beam will be stressed about 14 days after casting, the concrete strength at transfer will be approximately 40 N/mm^2. The allowable transfer stresses will therefore be $f_{ct_{adt}} = 2.55 \text{ N/mm}^2$, and $f_{cc_{adt}} = -20 \text{ N/mm}^2$.

From eqs. (5.26), at transfer

$$f_1 = 3.43 \text{ N/mm}^2 \quad \text{and} \quad f_2 = -18.53 \text{ N/mm}^2$$

so that the section is overstressed in Face 1 (upper face). One method of decreasing the tensile stress is to increase the prestress force. From eq. (5.10), $P_{max} = -773 \text{ kN}$, therefore increase the effective prestress force to 600 kN.

From eqs. (5.13),

(a) $e \geqslant 152 \text{ mm}$
(b) $e \leqslant 236 \text{ mm}$
(c) $e \geqslant 185 \text{ mm}$
(d) $e \leqslant 228 \text{ mm}$

so that, for an effective prestress force of 600 kN, $185 \text{ mm} \leqslant e \leqslant 228 \text{ mm}$. Therefore place the tendon at an eccentricity of 200 mm. (Note that with the increase in prestressing force, the critical equations (5.13c) and (5.13d) now provide the limiting condition.) The initial prestress force is therefore 750 kN (allowing for 20 per cent losses), and the characteristic strength required is at least 1071 kN. Therefore use one 40 mm dia prestressing bar (characteristic strength = 1250 kN).

At point of maximum moment, the effective depth = 500 mm. Initial strain $\epsilon_p = 0.0024$. From Section 5.5, the neutral axis depth is approximately 265 mm and the tendon stress at failure 0.752 kN/mm^2. From eq. (5.21),

$$M_u = 0.752 \times 1257(500 - 120) = 359 \text{ kN m}$$

which is acceptable.

The initial prestressing force is 750 kN m, so that, from eqs. (5.26),

$$f_1 = 1.55 \text{ N/mm}^2 \quad \text{and} \quad f_2 = -19.6 \text{ N/mm}^2$$

which is just acceptable. The tendon profile should now be checked for the transfer conditions, to ensure that the transfer stresses are nowhere exceeded.

The remainder of the calculation is concerned with the shear conditions and the detail design. At the centre of the span the shear is a minimum (theoretically zero) so that no shear reinforcement is necessary.

The above example illustrates the difficulties that may arise in the design of a simple prestressed section, and which result from the necessity of satisfying the various stress and limit state conditions. The reader may care to repeat the design for a reinforced concrete section, when it will be found that a section some 25 per cent heavier is required, thus demonstrating the advantage to be gained from prestressed construction.

Additional reading

BATE, S.C.C. and BENNETT, E.W. (1976) *Design of Prestressed Concrete*. London: Surrey University Press/Intertext.

BENNETT, E.W. (1973) *Structural Concrete Elements*. London: Chapman & Hall.

GUYON, Y. (1972/4) *Limit State Design of Prestressed Concrete*, vols 1 and 2. Toronto: Applied Science.

KONG, F.K. and EVANS, R.H. (1987) *Reinforced and Prestressed Concrete*. London: Van Nostrand Reinhold.

WILBY, C.B. (1981) *Post tensioned prestressed concrete*. Toronto: Applied Science.

6 Composite Construction

6.1 Introduction

The dictionary definition of the word 'composite' is 'made up of distinct parts', so that in reality any structure may be considered as being a composite structure, as virtually all completed structures involve two or more parts. Notice that in this definition no mention is made of materials, so that there is no prerequisite that the distinct parts must be of different materials. In structural/civil engineering, however, the term 'composite construction' is usually taken to mean a construction that involves a combination of concrete and steel sections in which both materials are load-carrying, although it may also be used to describe other material combinations. The use of a concrete covering merely to provide fire protection to a steel framed structure does not therefore constitute a composite structure in the accepted sense of the term, although it would satisfy the dictionary definition. Composite construction that involves the use of other materials is usually specified in terms of the materials used, e.g. timber/plastic composite or timber/steel composite.

Although the concrete/steel section combination is the most common form of composite structure, the principles of composite analysis and design are also frequently used in structures involving a combination of precast concrete units together with in-situ concrete. An example of this is given in Example 6.2, where it may be seen that the procedure of analysis is quite straightforward. This chapter will concentrate on introducing the basic analysis and design procedures for conventional concrete/steel composite structures.

Before commencing a detailed consideration of the composite construction technique, it is of interest to ask the question: why use composite construction? The two principal structural materials of steel and concrete each have their respective advantages and disadvantages, and it is rare to find a situation in which one material is an automatic choice for the particular job in hand. The use of composite construction goes some way towards combining the speed of structural erection provided by steelwork with the flexibility of concrete and its suitability for covering large areas

153

at a comparatively reasonable cost, and leads to the most common form of composite construction in which a reinforced concrete slab spans between steel beams.

This form of structure is found mainly in road bridges and multi-storey structures, and utilises the condition that if the connection between beam and slab is able to transmit the longitudinal shear forces then the section may be considered to be a homogeneous T section in which the slab acts as the compression flange. Transmission of the longitudinal shear forces is achieved either by means of shear connectors, which take many forms, but typically are steel studs welded to the flange of the beam, or by completely enclosing the steel beam in concrete so that outwardly the section looks like a simple concrete T beam. In the latter case the concrete enclosing the steel section is disregarded in the design, and serves only to ensure adequate connection between the concrete slab and the steel beam. Examples of the more common forms of shear connector are shown in Fig. 6.1.

The use of composite construction can lead to considerable economy, particularly in the amounts of steel used when compared with other forms of construction. In the early 1960s a comparative investigation made jointly by the Ministry of Public Building and Works and the British Constructional Steelwork Association on the construction of a ten-storey office block showed that approximately 20 per cent reduction in the overall costs of the building structure was obtainable by using composite construction, and when all costs are considered, similar economies may still be obtained today.

Design of a composite section requires knowledge of both concrete and steelwork design procedures, since, although the composite section is assumed to resist bending and deflection, the steel beam must also be checked for the various conditions of web failure (bearing, buckling and shear) that may occur. The methods of analysis and design are basically the same as those that have already been discussed in earlier chapters, and either the modular ratio or the limit state approach may be used.

One area in which composite sections differ from the more conventional reinforced concrete sections is that of self-weight. It is usual to attach the formwork or shuttering for the concrete directly to the steel beam, so that, when the concrete is placed, all of the concrete self-weight is carried by the steel sections. The application of live loads or further dead loads after the concrete has set results in stresses that are carried by the composite section, but the steel sections are already stressed by the wet concrete load, so that the critical stress may be developed in the steel at a load rather less than would be expected. This leads to inefficiencies in that not every element of the composite section is necessarily working to its maximum capacity, but the effect may be reduced or even eliminated entirely by supporting the steel sections while the concrete is being

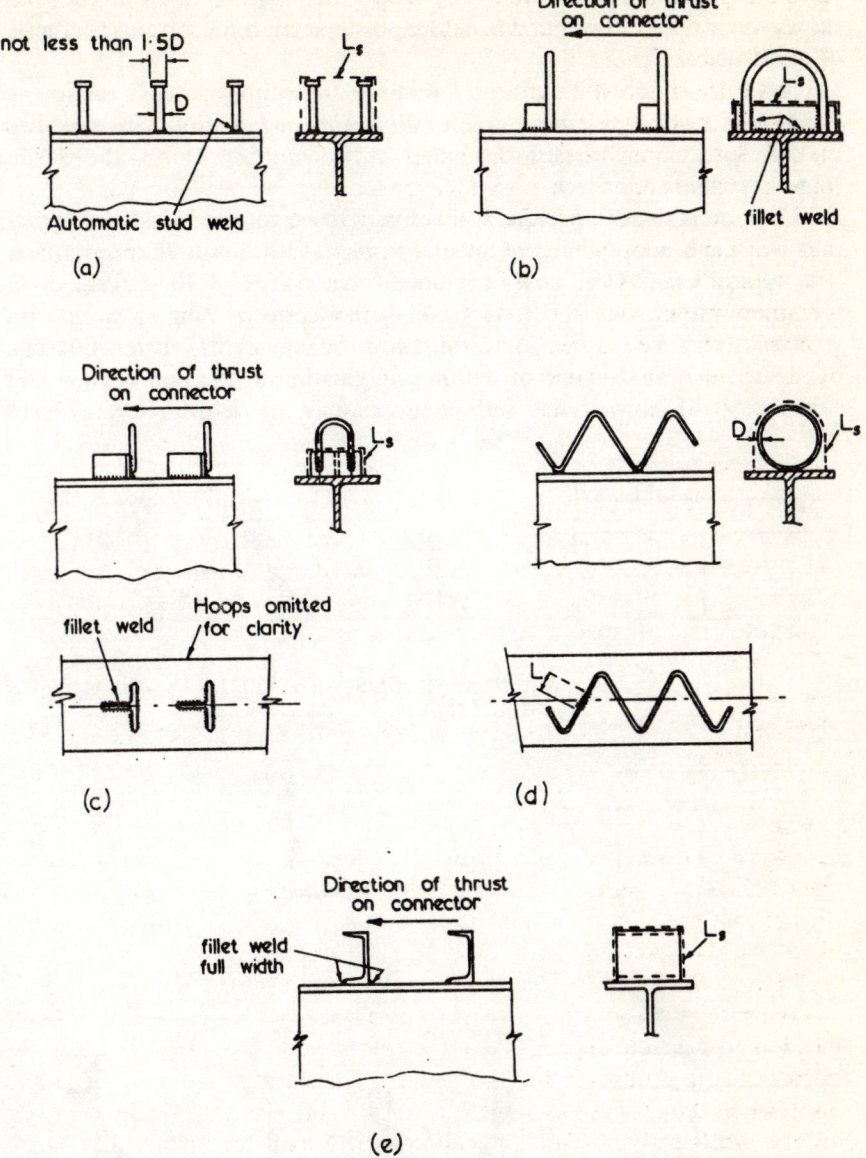

Fig. 6.1 Common forms of shear connector. (a) Stud connector. (b) Bar connector. (c) T connector. (d) Helical connector. (e) Channel connector.

poured. This is usually achieved by means of props or jacks. Figure 6.2 shows the stresses developed in a composite section for various methods of construction.

The code of practice currently relating to composite construction is CP 117: Part 1: 1965 *Simply supported beams in building*. This specifies that the section may be designed either by the load factor or by the elastic (modular ratio) approach.

This code is under revision, and is likely to be reissued in several parts that will each adopt the general philosophy of the limit state approach first used in CP 110 and now developed further in BS 8110, and the code for structural steelwork, BS 5650. The new code relating to composite structures has been issued in a draft form for comment, but reaction has been such that at the time of writing, publication of the new code is still some way off, and it has not been possible to determine the exact

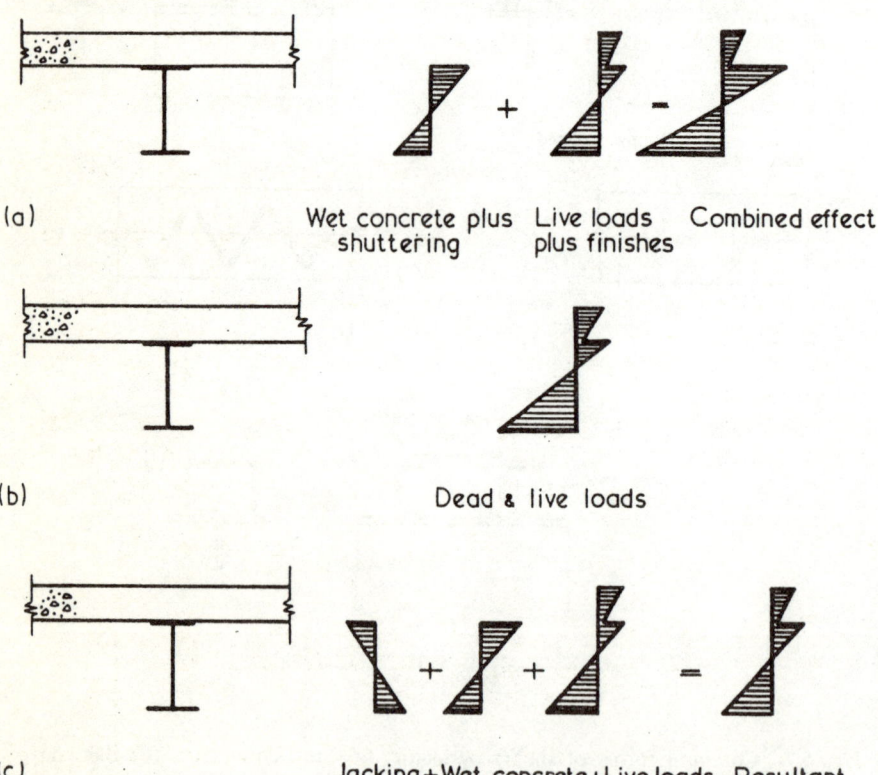

(a)

Wet concrete plus   Live loads   Combined effect
    shuttering   plus finishes

(b)

Dead & live loads

(c)

Jacking+Wet concrete+Live loads = Resultant

Fig. 6.2 Effects of different procedures of construction on the elastic stresses in the composite section. (a) No props. (b) Lower flange continuously propped. (c) Lower flange jacked.

recommendations that will be incorporated into the new standard. Any comments regarding the code of practice must therefore be taken as relating to the 1965 edition of CP 117.

Although the design and analysis of composite sections may be performed using either load factor or elastic methods, if load factor procedures are used then checks must be made to ensure that the stresses obtained by an elastic analysis do not exceed certain proportions of the specified steel yield and concrete cube stresses. This ensures that the serviceability limit state is not exceeded, although the code of practice does not refer to the condition as such. While the revised CP 117 will make specific references to the serviceability limit state, the method of complying with its requirements will not differ greatly from that in the existing code.

6.2 Elastic analysis and design

The design procedure for the elastic condition follows that which would be used in the elastic design of a T beam, except that the analysis of the section must be carried out by the method of 'transformed sections'. This was discussed in Chapter 2, and involves considering the steel in terms of equivalent concrete units (or alternatively considering the concrete in terms of equivalent steel units).

The neutral axis position is first determined. A convenient method is to take first moments of equivalent area about a suitable axis. The NA may lie in one of two areas: (i) in the concrete slab, or (ii) in the steel section. If it is assumed that the neutral axis passes through the equivalent centroid of the composite section, then the most convenient axis about which to take moments lies along the top of the concrete slab.

Consider the section shown in Fig. 6.3. The effective width of the slab is obtained from the same considerations as govern the effective width of

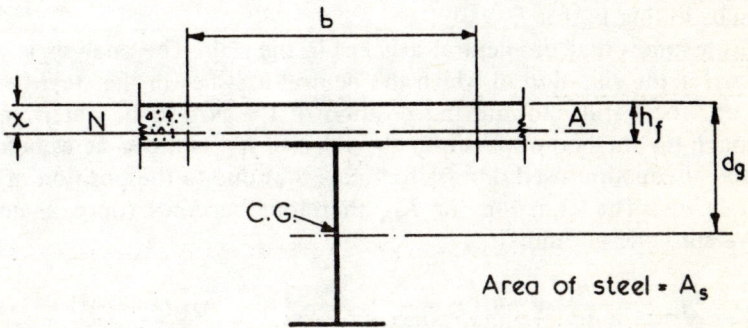

Fig. 6.3

reinforced concrete T beams designed by the modular ratio method as shown in Chapter 2, and taking area moments in equivalent concrete units about the top of the slab gives

$$\frac{bh_f^2}{2} + A_s \alpha_e d_g = (bh_f + A_s \alpha_e)x \tag{6.1}$$

which allows the position of the neutral axis to be obtained. This applies to all positions of the NA. The second moment of area I_{NA} (about the neutral axis) is obtained, once again working in equivalent concrete units, by

$$I_{NA} = \frac{bx^3}{12} + \frac{bx^3}{4} + \alpha_e[I_s + A_s(d_g - x)^2]$$

$$= \frac{bx^3}{3} + \alpha_e[I_s + A_s(d_g - x)^2] \tag{6.2}$$

Suppose that the section is subjected to a sagging moment M. The extreme fibre stresses f_c and f_t are given by

$$\frac{f_c}{x} = \frac{M}{I_{NA}} \quad \text{and} \quad \frac{f_t}{(h-x)} = \frac{M\alpha_e}{I_{NA}} \tag{6.3}$$

the α_e factor being necessary to change from concrete units to steel units. Hence the ratio f_c/f_t may be evaluated. This ratio may then be compared with the ratio p_{cb}/p_{st} to determine whether the section is over- or under-reinforced; if the stresses are increased at a uniform rate (so that f_c/f_t remains constant), it may be determined which material will reach its permissible stress value first. If (as is usual) $p_{cb}/p_{st} > f_c/f_t$, the section is under-reinforced and the limiting stress will be reached in the steel beam. The permissible bending moment is then obtained from the equation

$$\frac{f_t}{(h-x)} = \frac{M\alpha_e}{I_{NA}} \tag{6.3}$$

but substituting p_{st} for f_t.

This assumes that the neutral axis lies in the slab. The analysis is very similar for the situation in which the neutral axis lies in the steel beam, with eq. (6.1) still allowing the position of the NA to be determined. Although the method of obtaining the value of I_{NA} is the same as before, the actual equation used differs from eq. (6.2) due to the position of the neutral axis. The equation for I_{NA} therefore becomes (once again in equivalent concrete units)

$$I_{NA} = \frac{bh_f^3}{12} + bh_f\left(x - \frac{h_f}{2}\right)^2 + \alpha_e[I_s + A_s(d_g - x)^2] \tag{6.4}$$

and the remainder of the calculation follows that for the situation in which the NA is in the slab.

It is more usual to use the load factor method for design of composite sections, but in order to prevent excessive and permanent deformations under working load it is necessary to check the elastic stresses at working load, and to ensure that the maximum stresses do not exceed $0.9 \times$ yield stress in the steel and 0.333 of the cube strength in the concrete. This is the check of the serviceability condition referred to earlier.

6.3 Load factor design

The equations that will be developed here for the load factor method of analysis are based on the same assumptions as those used for the design of reinforced concrete sections (Chapter 3). As mentioned above, the standard relating to the use of composite sections is under revision, and when it is eventually published, the equations and recommendations which it contains may not exactly match the equations that follow. However, the general approach is expected to be very similar, so that the student of composite construction will have little difficulty in understanding any variations in the design equations.
The assumptions used are:

1. The whole of the steel section is stressed to a yield value whether it be in tension or in compression. The same load factor is applied to both tensile and compressive zones.
2. Concrete below the neutral axis is unstressed, i.e. tensile stresses in the concrete are ignored.
3. The rectangular stress block for concrete in compression is assumed, with the compressive stress being $0.45f_{cu}$ over a depth of $0.9x$ (x being the depth to the neutral axis) for sections where the NA falls within

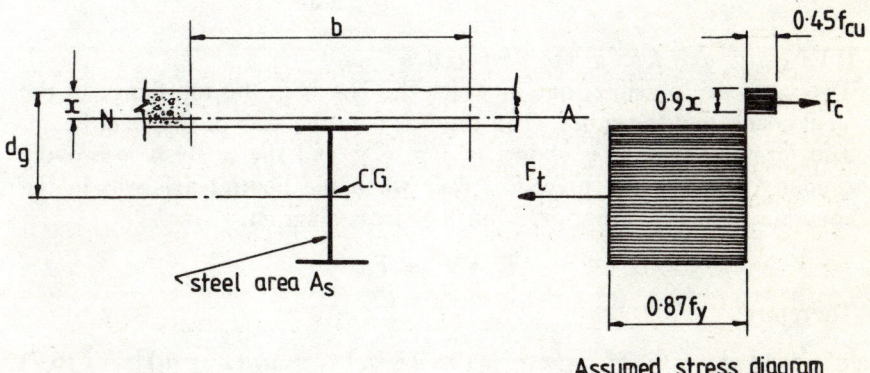

Fig. 6.4 Composite section having the neutral axis in the concrete slab.

the concrete slab, or $0.4f_{cu}$ over the full concrete depth for sections where the NA lies in the steel beam.

Calculation of the effective breadth is made by the same procedure as used for load factor design of reinforced beams, which as shown in Chapter 3, leads to the recommendation that the effective breadth shall not extend more than span/10 on either side of the steel beam web.

As with the elastic analysis, two conditions may arise: (a) the neutral axis falls in the slab; (b) the neutral axis falls in the steel section.

(a) Neutral axis in the concrete slab

Consider the system shown in Fig. 6.4. Equating forces on the stress diagram,

$$F_c = F_t$$

Therefore

$$0.4f_{cu}bx = 0.87A_sf_y$$

giving

$$x = \frac{2.17A_sf_y}{f_{cu}b} \qquad (6.5)$$

Taking moments about F_c gives the ultimate moment

$$M_U = F_t z$$

but

$$z = (d_g - 0.45x)$$

therefore

$$M_U = 0.87A_sf_y\left(d_g - \frac{0.98A_sf_y}{f_{cu}b}\right) \qquad (6.6)$$

(b) Neutral axis falls in the steel section

Two cases are possible, one in which the NA is in the top flange of the steel beam, and the other when the NA is in the web of the steel beam. The former situation is shown in Fig. 6.5, and the analysis is broadly similar to that of the preceding case when the neutral axis was in the concrete slab. Equating forces on the stress diagram,

$$F_t = F_c + F_{sc}$$

Therefore

$$0.4f_{cu}h_f + 0.87f_yb_{sf}(x - h_f) = 0.87f_y[A_s - b_{sf}(x - h_f)] \qquad (6.7)$$

hence the depth to the neutral axis may be found.

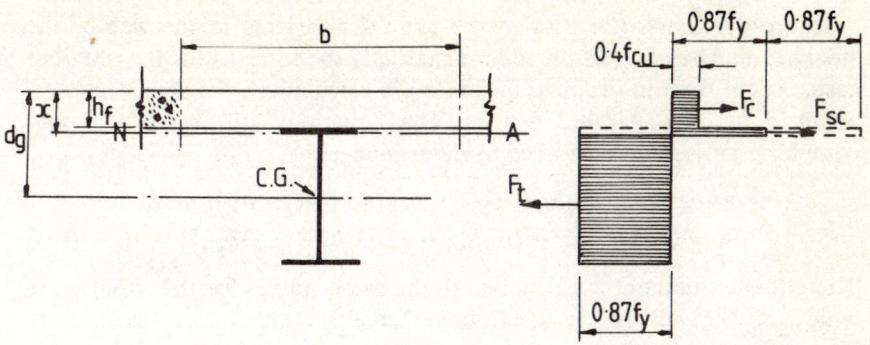

(b) Assumed stress diagram

Fig. 6.5 Composite section having the neutral axis in the steel flange.

The ultimate moment is again obtained by taking moments about the line of action of F_c. However, if F_{sc} is added to each side of the stress diagram, the arithmetical calculation is considerably simplified as the line of action of the tensile force may then be taken as being through the centroid of the steel. Figure 6.5(b) shows the stress diagram after adding F_{sc} to both sides, and taking moments about F_c gives

$$0.87A_sf_y\left(d_g - \frac{h_f}{2}\right) - 2 \times 0.87f_y[b_{sf}(x - h_f)]\frac{x}{2} = M_U \qquad (6.8)$$

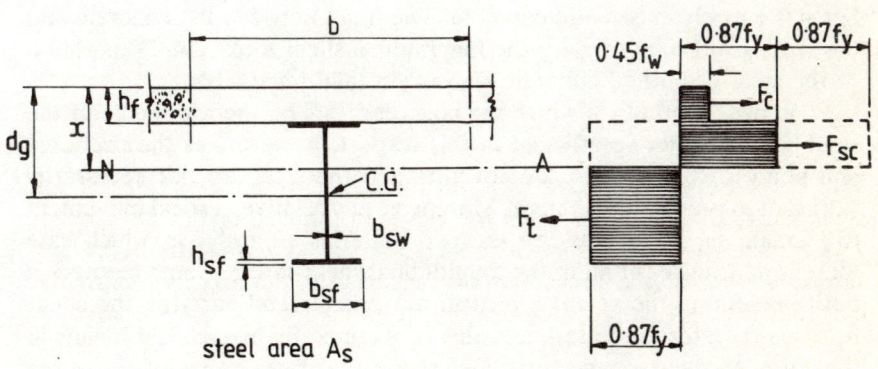

(b) Assumed stress diagram

Fig. 6.6 Composite section having the neutral axis in the steel web.

Figure 6.6 shows the case of the neutral axis lying in the web of the steel beam. The analysis procedure is exactly the same as for the previous case, except that, in order to facilitate the calculation, the steel section is assumed to have parallel flanges. The equations are, however, more complex, e.g., equating forces to determine x:

$$0.4f_{cu}bh_f + b_{sf}h_{sf}0.87f_y + b_{sw}(x - h_f - h_{sf})0.87f_y$$
$$= 0.87f_y\{A_s - [b_{sf}h_{sf} + (x - h_f - h_{sf})b_{sw}]\} \qquad (6.9)$$

The ultimate moment is calculated in the same way as for the situation in which the NA falls in the flange of the steel section.

6.4 Shear connectors

The application of a bending moment produces longitudinal shear stresses in a section. If the section is made up of materials with differing moduli of elasticity, this results in differential strains being developed, and movement of one material of the section relative to the other, so that the section ceases to act as a homogeneous whole and failure may occur.

In reinforced concrete sections the shear forces between concrete and reinforcement are resisted by the bond that exists between the two materials. In such cases the problem is simplified in that, since it is completely surrounded by concrete, the reinforcement cannot move in a vertical plane relative to the concrete (a movement that would occur if differential longitudinal strains were permitted, assuming only the ends to be restrained).

A similar situation exists in composite construction where the steel beam is entirely encased in concrete. The bond between the concrete and the steel is sufficient to carry the longitudinal shear forces, and separation of the steel beam and concrete slab is virtually impossible.

Composite sections in which the concrete rests on the top flange of the steel beam are not so efficient in this respect, however, as the frictional and bond forces between the concrete and the steel are not necessarily sufficient to prevent movement. Moreover, any relative vertical movement will break the bond between the two materials entirely, in which case there is no transfer at all of the longitudinal shear forces. Some method of both preventing the relative vertical movements and carrying the shear forces is therefore required, and this is obtained by mechanical means in the form of shear connectors. Figure 6.1 illustrates some of the more common types of shear connector; the essential requirement is that the connector be able to carry both horizontal and vertical forces. The connector is welded to the top flange of the steel section and the concrete then poured around it.

6.4.1 Design of shear connectors

Unlike the composite section, which may be designed by either load factor or elastic procedures, shear connectors must be designed by load factor methods only. The shear connectors are taken as providing all of the resistance to the longitudinal shear force between the steel beam and the concrete slab, no allowance being made for any bond or frictional forces developed between the two materials. The number of connectors provided should therefore be sufficient to resist the total maximum value of the longitudinal shear forces; the basic calculation is theoretically quite simple, and merely involves calculating the maximum longitudinal shear force between the concrete and the steel, and dividing this by the design load of each connector to establish the number of connectors required.

The design load of a connector depends on a number of factors: the size and type of the connector, the strength of the concrete, and the type of loading that is envisaged. This latter effect is particularly important in the case of composite bridges, where the cycle of loading may cause problems of fatigue. The size of the connector is governed to a large extent by the requirement that the connector be able to prevent vertical separation of the slab and beam. This is achieved by specifying a minimum penetration of the connector into the slab of not less than 50 mm. In addition, CP 117 specifies that the connector shall project not less than 25 mm into the compression zone of the concrete slab. This ensures that the head of the connector is enclosed by a volume of uncracked concrete, whereas if the requirement were simply that the connector was to project 50 mm into the slab, the head of the connector might still be in the cracked tensile zone of the concrete, and might be able to pull out of the slab together with the surrounding concrete block, thus failing to fulfil the holding-down requirement. The holding-down performance of a stud connector is also enhanced by the specification that the diameter of the head of a stud connector be not less than $1 \cdot 5 \times$ the diameter of the body of the stud.

Having established the number of connectors required, their spacing may be determined. For a uniform loading system the connectors are uniformly spaced between the end of the beam and the point of maximum bending moment, but for heavy concentrated loads the distribution should be varied according to the relative areas of the shear force diagram between the points of shear discontinuity, although the spacing may be constant over the portion of the beam between the shear discontinuities. The choice of spacing is also subject to certain other constraints, since the connectors must be neither so close together that compaction of the concrete between them cannot easily be carried out, nor so far apart as to give rise to the possibility of local slab deformations due to separation of the slab and the beam. This latter point is covered by specifying that the spacing of connectors should not be greater than four times the slab thickness or 600 mm, whichever is the less.

Assuming that the sizes of the concrete slab and the steel beam have been determined, the design of the connectors may be summarised as follows:

1. Determine the longitudinal shear force between the concrete slab and the steel beam. This may be evaluated from the compressive forces acting in the concrete, as shown by the stress diagram.
2. Choose a suitable connector type and size. The most common connectors are stud connectors, since these are usually the most economical to fix onto the steel beam. The size of the connectors is influenced by the holding-down requirements.
3. Determine the spacing of the connectors by first obtaining the number required, and then spacing these equidistantly along the relevant part of the steel section.
4. Check that the spacing thus obtained satisfies the empirical spacing outlined above. If it does not, a different sized connector may be required.

Even though the provision of suitable connectors ensures that no longitudinal separation of the slab and beam occurs, there is still the possibility that the longitudinal shear stress in the concrete that surrounds the connectors is excessively high and that shear failure may occur in the concrete on a plane that is close to the beam/slab interface. This is of particular importance in the case of haunched beams, and the code of practice makes special reference to beams of this form. For other, non-haunched beams, however, the possibility of longitudinal shear failure within the concrete is covered by restricting the maximum shear force in the concrete to certain limits, and ensuring that a minimum area of transverse reinforcement is provided in the bottom of the slab. The limits placed by CP 117 on the longitudinal shear force in the concrete are that the shear force should not exceed the lesser of

$$0.24L_s(f_{cu})^{1/2} + A_sf_yn \qquad (6.10a)$$

$$0.64L_s(f_{cu})^{1/2} \qquad (6.10b)$$

where L_s is the smaller of either the length of the shear surface around the connector or connector group, or twice the slab thickness (for a T beam), A_s is the area of transverse reinforcement in the bottom of the slab (per mm length of beam), n is the number of times that the transverse reinforcement is intersected by the shear surface ($n = 2$ for T beams), and f_{cu} and f_y are the material strengths. The area of transverse reinforcement A_s is given by

$$A_s = \frac{V}{4f_y} \qquad (6.11)$$

where V is the shear force per mm run of the beam.

More recent research has shown that the recommendations of equations 6.10 and 6.11 are unduly conservative, so that the new code of practice will present alternative equations for the provision of transverse reinforcement.

Other factors which need to be considered in the design of composite sections are creep and shrinkage of the concrete and the effects of temperature variation between the concrete and the steel. Detailed consideration of these effects involves a deeper understanding of the subject than may be gained from this text; the reader is referred to the additional material listed at the end of the chapter. Briefly, the effects are as follows: both creep and shrinkage affect the composite section in similar fashion, causing stress to be transferred from the concrete to the steel. Evaluation of this stress transfer is difficult and several methods of varying complexity have been suggested. Because of the complicated nature of exact calculations, most of the solutions adopt approximations of varying degrees, but the effects of stress transfer due to both creep and shrinkage are frequently so small that they can be ignored for practical purposes. Certainly, CP 117: Part 1: 1965 makes no direct reference to creep and shrinkage effects, and Part 2: 1967 of the same code recommends that the effects may be ignored in simply supported beams with shear connectors.

The thermal effects are of more importance in bridge structures than in most multi-storey construction (except perhaps for the roof slabs), and arise from the temperature differences that occur through the exposure of the slab to direct solar heat, while the steel sections remain in the shade of the slab. This effect obviously varies with the location of the structure, but the code recommends that a temperature difference of $\pm 10°C$ be used for structures in the United Kingdom.

Example 6.1

As an example of the calculations of a composite structure, consider a beam and slab floor in which the beams are at 3 m centres and are taken as simply supported over a span of 8 m. Assuming that the premises are to be used for office accommodation with provision for filing and storage space, the live load intensity will be taken as being 5 kN/m². Other dimensions, material properties and loading details are as listed below, and it is further assumed that the beams are not propped during construction, and that the concrete slab has already been designed.

Slab thickness $= 0.15$ m, allowance for finishes $= 1$ kN/m², $f_{cu} = 30$ N/mm², $f_y = 250$ N/mm², $\alpha_e = 15$, and the density of reinforced concrete is 2400 kg/m³.

Elastic design. The bending moment due to the wet concrete will be taken by the steel beam alone, whereas the bending moment due to the live load plus the finishes will be taken by the composite section.

Bending moment due to the wet concrete is

$$\frac{\omega l^2}{8} = \frac{3 \times 0\cdot15 \times 2400 \times 64 \times 9\cdot81}{8} = 84\cdot8 \text{ kN m}$$

plus some allowance for the self-weight of the steel beam (say 60 kg/m),

$$\frac{60 \times 64 \times 9\cdot81}{8} = 4\cdot71 \text{ kN m}$$

Therefore total bending moment on the beam alone, $M_1 = 89\cdot5$ kN m.

Bending moment due to the live load plus the finishes

$$M_2 = \frac{6 \times 3 \times 64}{8} = 144 \text{ kN m}$$

Therefore the total bending moment $= M_1 + M_2 = 233\cdot5$ kN m.

Some indication of the beam size required may be obtained by considering the section that would be needed if the composite action were not effective. In that case, assuming that a beam of Grade 43 specification was used, the required Z value would be

$$Z = \frac{233\cdot5 \times 10^6}{165} \text{ mm}^3 = 1415 \text{ cm}^3$$

Therefore try a $457 \times 152 \times 52$ UB section ($Z = 949$ cm^3): the stress in the beam due to wet concrete and self-weight $= M_1/Z = 90\cdot5$ N/mm^2. This is less than the permissible stress for steel in bending and is therefore acceptable.

The cross-section of the composite section is therefore as shown in Fig. 6.7, with the effective width of the concrete flange being the least of span/3, distance between the beams, or beam web width plus (12 × slab thickness) (see Chapter 2).

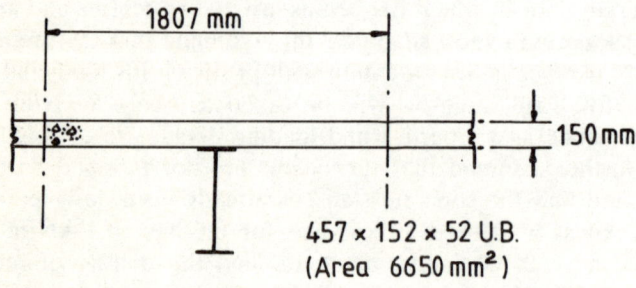

1807 mm

150 mm

457 × 152 × 52 U.B.
(Area 6650 mm²)

Fig. 6.7

Taking first moments of area about the top of the slab,

$$x[(6650 \times 15) + (1807 \times 150)] = \left(\frac{1807}{2} \times 150^2\right) + (6650 \times 15 \times 374 \cdot 9)$$

Therefore $x = 156$ mm (i.e. the neutral axis is in the steel section).
From eq. (6.4),

$$I_{NA} = 1 \cdot 03 \times 10^6 \text{ cm}^4 \quad \text{(in equivalent concrete units)}$$

From eqs. (6.3), maximum stress in steel section due to the live load and finishes is given by

$$
\begin{aligned}
f_t &= \frac{M_2 \alpha_e (h - x)}{I_{NA}} \\
&= \frac{144 \times 10^6 \times 15(599 \cdot 8 - 156)}{1 \cdot 027 \times 10^{10}} \\
&= 93 \cdot 3 \text{ N/mm}^2
\end{aligned}
$$

and the maximum stress in the concrete by $f_c = (M_2 x)/I_{NA} = 2 \cdot 19$ N/mm^2.

$$\frac{f_c}{f_t} = \frac{2 \cdot 19}{93 \cdot 3} = 0 \cdot 023$$

$$\frac{p_{cb}}{p_{st}} = \frac{10}{165} = 0 \cdot 06$$

Therefore

$$\frac{p_{cb}}{p_{st}} > \frac{f_c}{f_t}$$

so that the section is under-reinforced, i.e. the limiting stress will be reached in the steel. The permissible moment of resistance of the composite section therefore is given by

$$MR = \frac{p_{st} I_{NA}}{(h - x) \alpha_e} = 255 \text{ kN m}$$

The basic elastic design is now complete, although checks on the strength of the web of the steel beam and calculation of the shear connector requirements remain to be made. Before going further with the design, however, the total stress-distribution in the section must be checked.

The total stress diagram (Fig. 6.8) shows that the maximum stress in the steel section is 182·6 N/mm^2. This is greater than the permissible stress of 165 N/mm^2 for a Grade 43C material, so that the section chosen is not sufficiently strong, and another slightly larger beam section should be used. Alternatively, since the concrete stress of 2·19 N/mm^2 is considerably lower than that permitted for this grade of concrete, propping

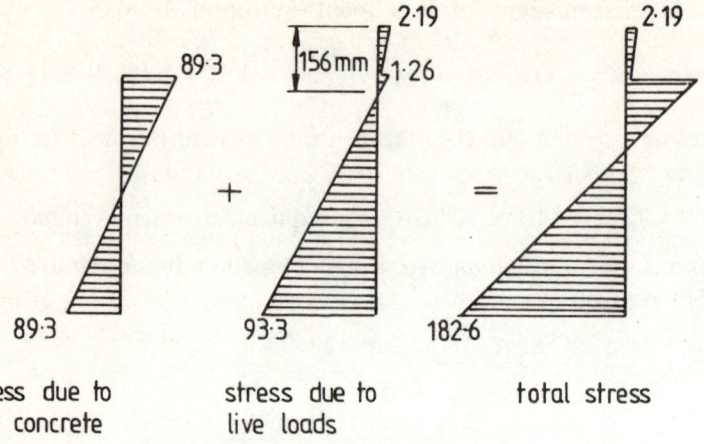

stress due to   stress due to   total stress
wet concrete   live loads

Fig. 6.8 Stresses (N/mm^2) in the section of Example 6.1.

or prior jacking of the steel beam could be employed. This would have the effect of increasing the stresses in the concrete slab while reducing the stresses on the steel beam due to the wet concrete.

Recalculate the design using the load factor method. Using load factors of 1·4 on dead loads, and 1·6 on live loads gives $M_u = 1·4M_1 + 1·6M_2 = 355·7$ kN m. Try the same section size (457 × 152 × 52 UB) to Grade 43. The cross-sectional form of the composite structure is therefore the same as that used in the previous calculation, but because of the different assumptions made in defining the effective width, the effective width of the composite section designed by the load factor method is 1607 mm. Assuming that the neutral axis lies in the slab, from eq. (6.5)

$$x = \frac{2·17 \times 6650 \times 250}{30 \times 1607} = 75 \text{ mm}$$

The assumption is therefore correct, and eq. (6.6) may be used to determine M_u:

$$M_u = 0·87 \times 6650 \times 250\left(374·9 - \frac{0·98 \times 6650 \times 250}{30 \times 1607}\right) = 493 \text{ kN m}$$

which is greater than the required M_u and is therefore acceptable.

Check the serviceability conditions of maximum stresses in the concrete and steel under working loads. This condition is satisfied if $f_t \ngtr 0·9f_y$ and $f_c \ngtr f_{cu}/3$. From the elastic design,

$$f_t = 182·6 \text{ N/mm}^2 \ngtr 0·9f_y$$

and

$$f_c = 2 \cdot 19 \text{ N/mm}^2 \not{>} \frac{f_{cu}}{3}$$

Therefore this section is acceptable.

As with the elastic design, the steel beam now requires checking for the buckling, bearing and shear forces at the supports. However, ignoring such considerations, had a UB section been required to carry the whole of the load system without composite action, a $457 \times 152 \times 67$ section would have been required. In this example, therefore, a weight saving of steel of approximately 30 per cent has been achieved, although the addition of shear connectors will reduce this saving.

Calculation of shear connector requirement. The ultimate compression force in the slab $= F_c$ where

$$F_c = 0 \cdot 4 f_{cu} x b$$

(x being replaced by h_f for situations where $x > h_f$)

$$F_c = 0 \cdot 4 \times 30 \times 74 \times 1607 = 1 \cdot 43 \text{ MN}$$

This force varies linearly from zero at the ends of the beam to a maximum at the centre of the span.

Values for stud connectors tabulated in CP 117 show that the design strength of a 19 mm dia $\times$ 75 mm high connector in concrete having a strength of 30 N/mm^2 is approximately 76 kN per stud. The number of connectors required is therefore $(1 \cdot 43 \times 10^3)/76 = 18 \cdot 8$, i.e. nineteen connectors spaced equidistantly over half the span of the beam: $4/19 = 0 \cdot 211$ m spacing. However, this is not a convenient spacing, and since the spacing cannot be increased (giving less connectors than required) it must be reduced.

Therefore use 19 mm diameter $\times$ 75 mm high stud connectors at 200 mm centres. This spacing is less than the maxima laid down in the code of practice and is sufficiently large to give good compaction, so is therefore suitable.

Check of shear forces in the concrete. Shear force V per mm run of beam is calculated as

$$\frac{\left(\begin{array}{c}\text{load on one connector} \\ \text{at failure}\end{array}\right) \times \left(\begin{array}{c}\text{no. of connectors at} \\ \text{a cross-section}\end{array}\right)}{\text{longitudinal spacing of the connectors}}$$

$$= \frac{1 \cdot 43 \times 10^6 \times 1}{20 \times 200} = 358 \text{ N/mm}$$

The minimum area of transverse reinforcement required in the slab is given by eq. (6.11):

$$A_s = \frac{358}{4 \times 250} = 0.36 \text{ mm}^2/\text{mm run}$$

10 mm dia bars at 200 mm centres give $0.39 \text{ mm}^2/\text{mm}$ run, so use this as the minimum permissible transverse reinforcement.

The limits of permissible longitudinal shear force on the section are given by eqs. (6.10), which show that the longitudinal shear force must not exceed either

(a) $(0.24 \times 178.5 \times \sqrt{30}) + (0.39 \times 250 \times 2) = 430 \text{ N/mm}$

or (b) $0.64 \times 178.5 \times \sqrt{30} = 626 \text{ N/mm}$

The actual value of V is 358 N/mm, so these conditions are satisfied and the design may be considered complete.

Example 6.2

A beam and slab floor has been designed to carry a live load of 7.5 kN/m^2 over a span of 5 m. The beams are at 4 m centres and are simply supported. The designer has treated the section as a flanged beam and has produced the design shown in Fig. 6.9. The contractor proposes to construct the floor by placing precast beams (200×325 mm with two 32 mm dia bars) in position, and then constructing an in-situ concrete slab over the beams. Assuming that the joint between the beams and the slab is able to transmit shear, is this method of construction acceptable?

Assume $f_{cu} = 30 \text{ N/mm}^2$, $f_y = 460 \text{ N/mm}^2$, load factors of 1·6 (live load) and 1·4 (dead load), and reinforced concrete density of 2300 kg/m^3.

The example illustrates the care that must be taken in construction procedures. Using the proposed constructional sequence, the precast beams will carry *all* of the dead loads of the structure, whereas the live loads will be supported by the composite section.

For the precast beam,

$$d = 259 \text{ mm}$$

Dead load $= [(0.125 \times 4) + (0.2 \times 0.325)] \times 2300 \times 9.81 \text{ N/m}$
                   (slab)   (beams)

$$\doteqdot 12.75 \text{ kN/m}$$

Therefore the moment on the beam due to the dead load only

$$M_D = \frac{12.75 \times 5^2}{8} \times 1.4 = 55.8 \text{ kN m}$$

Analysis of the beam shows that a bending moment of 55·8 kN m requires

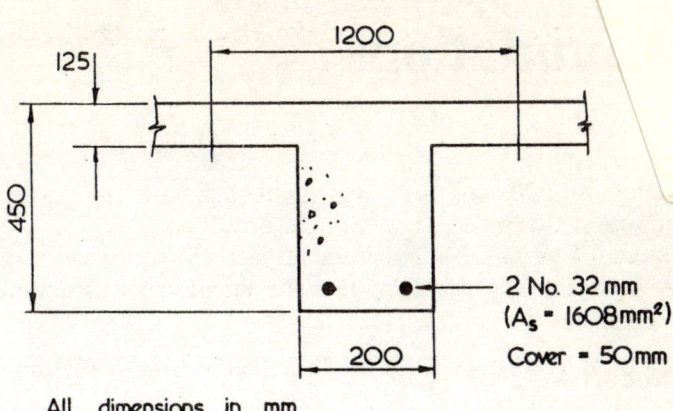

All dimensions in mm

Fig. 6.9

a neutral axis depth of 111 mm. The stress in the reinforcement will then be

$$\frac{55\cdot8 \times 10^6}{1608[259 - (111 \times 0\cdot45)]} = 166 \text{ N/mm}^2$$

The maximum live load moment = $7\cdot5 \times 4 \times (5^2/8) \times 1\cdot6 = 150$ kN m, and analysis of the T beam section gives $x = 28$ mm. The stress in the reinforcement due to the live loads only is therefore

$$\frac{150 \times 10^6}{1608[384 - (0\cdot45 \times 28)]} = 251 \text{ N/mm}^2$$

The *total* stress in the reinforcement is therefore 251 + 166 = 417 N/mm², which is greater than the factored yield stress and is therefore not acceptable.

A check on the ultimate moment capacity of the T beam shows that its ultimate moment of resistance is 211 kN m. The ultimate moment to be carried is 206 kN m, so that the original design is satisfactory, and the proposed method of construction requires modification. One suitable modification would be to support the precast units while the in-situ slab was being cast.

Additional reading

JOHNSON, R.P. (1982) *Composite Structures of Steel and Concrete*, vol. 1. London: Granada.

7 Foundations

7.1 Introduction

A detailed treatment of foundation design obviously involves extensive discussion of soil properties and the various methods of analysis used in soil mechanics, which is not appropriate to this text. Most foundations involve the use of concrete in some way, however, so that it is not entirely irrelevant to consider the design of various forms of simple foundation.

The main feature requiring consideration in foundation design is the bearing pressure that can be supported by the ground beneath the foundation. This determines the minimum plan area of the foundation. If the foundation carries an axial load, the pressure beneath the foundation is uniform and may be directly equated to the allowable maximum ground pressure; but if the load is placed eccentrically on the footing, or includes a bending moment transmitted through the column(s), the resulting pressure varies between a maximum at the edge nearest the line of action of the load and a minimum at the opposite edge; the allowable bearing stress on the ground must therefore reflect the maximum pressure occurring under the foundation.

A possible solution, though not always economic, is to make the foundation unsymmetrical in plan so that uniform pressure-distribution results from the eccentric load (Fig. 7.1). This has the advantage of saving on foundation material, but the additional complexity may increase the excavation and construction costs.

The simplest type of foundation is the isolated footing that carries the axial load from a single column. The minimum plan dimensions of the foundation are obviously restricted by the allowable bearing pressure, which also effectively determines the minimum thickness of the footing, since the thickness must be sufficient to allow the concentrated column loads to be distributed through the footing to give a uniform bearing pressure. The distribution is assumed to be independent of the material strength, and to occur at 45° (Fig. 7.2), which permits the foundation thickness to be calculated. The thickness obtained from this calculation

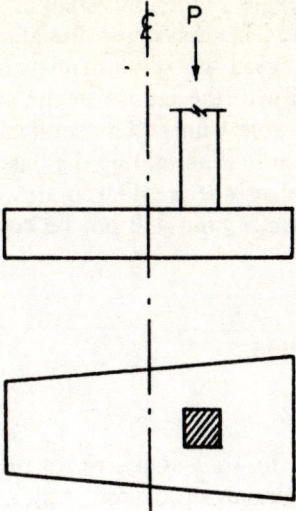

Fig. 7.1 Unsymmetrical foundation producing uniform ground pressure under an eccentric load.

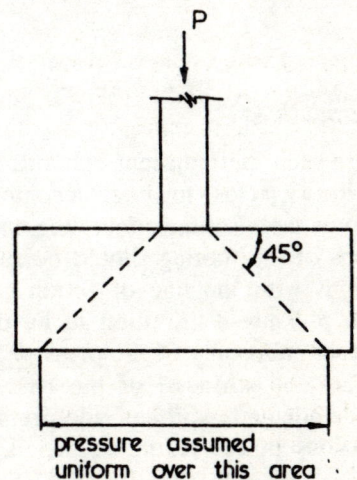

Fig. 7.2 Distribution of concentrated load through the foundation.

may not be sufficient, however, as checks of the shear stress, bending moment, and holding-down forces may require the depth to be increased.

Although the design of foundations with regard to shear forces and bending moments is carried out for the ultimate condition, in the calculation of the ground bearing pressures the limit state of serviceability is taken to be the critical limit state. This is because the allowable ground pressures are determined with regard to settlement, whereas ultimate ground pressures are concerned with the failure of the soil, and may show little relationship to the allowable values. The serviceability condition usually becomes the critical factor in determining the base area, and although the ultimate condition should also be considered, this involves a more detailed knowledge of soil mechanics, and will not be covered here.

7.2   Types of foundation

7.2.1   Isolated footings

A problem that is likely to arise as the result of using single or isolated footings is that of differential settlement. Although this can be allowed for in certain types of structure at the design stage, the design is thereby made more complex. In order to reduce the possibility of differential settlement, the area of isolated footings should ideally be adjusted in proportion to the column load, so that equal stress-distributions occur under each footing. ·

7.2.2   Combined foundations

If the column loads are such that adjacent columns require footings that either overlap or become very close together, a combined foundation may be used, with one, two or more columns founding on the same footing. If possible the dimensions of the footing should be such that the centroid of the foundation aligns with the line of action of the several loads. This allows a uniform pressure-distribution to be developed under the base, although the actual uniformity of the pressure also depends on the thickness, and therefore the stiffness, of the base. If the base is not thick enough to provide adequate stiffness, the pressures beneath it vary according to the positions of the columns carried, much as shown in Fig. 7.3.

In congested areas such as town centres and other areas of high building density, restrictions from adjacent structures may prevent the foundation from being centrally placed, resulting in high eccentricities giving in-admissible ground bearing pressures. In these conditions a balanced foundation may be achieved by providing a counterbalancing moment, either by tieing the column footing to that on the opposite side of the building or by constructing a mass concrete counterbalancing weight. A

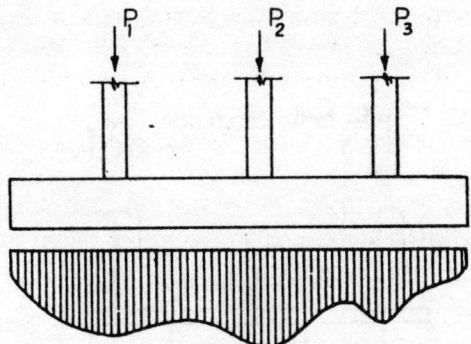

Fig. 7.3 Combined foundation. Insufficient stiffness of the foundation results in non-uniform ground pressure distribution, as shown in the hatched figure.

third alternative is to use some form of ground anchor such as tension piles. These alternatives are illustrated in Fig. 7.4.

7.2.3 Rafts

If the columns are closely spaced in one direction only, the combined foundation is extended to include the loads from all of the columns along that side of the structure; such a foundation is known as a strip foundation. When the columns are closely spaced in both directions and/or if the soil is particularly weak so that the permissible ground pressures are particularly low, a raft foundation covering the whole area of the structure is used. In general, raft foundations become economic if individual footings would occupy more than approximately 50 per cent of the ground area.

There are various forms of raft foundation, the most simple being the solid raft which is usually economic up to thicknesses of 0·5 m. The thickness is determined from consideration of the shearing forces and bending moments in the raft. If a thickness greater than about 0·5 m is required, a beam and slab type of construction may prove more economical. It is difficult to be specific about the thickness for which this form of construction becomes more economic than the solid raft, since the spacing of the columns (which are carried by the 'beam' components of the raft) obviously affects the amount of material that can be saved by the local reductions in thickness. The principle of the beam and slab raft is exactly the same as that of the beam and slab floor, the loads on the slab being obtained from the ground pressures (actual values, not permitted values) and the beam reactions being provided by the columns. As with the other forms of foundation, it is advantageous to make the centroid of the raft

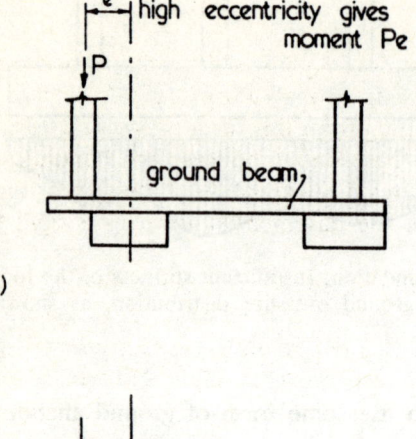

(a)

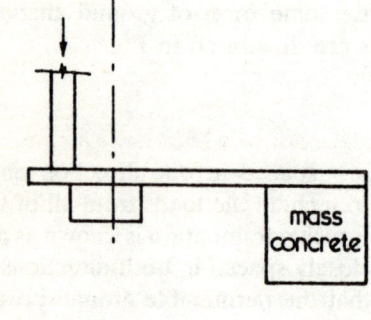

(b)

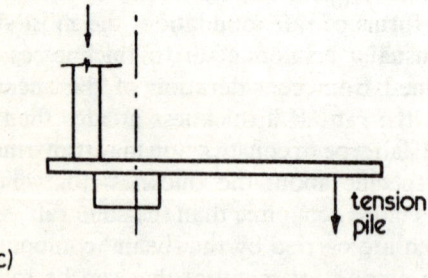

(c)

Fig. 7.4 Balanced foundations. (a) Counterbalancing moment provided by column on the opposite side of the structure. (b) Counterbalancing moment provided by mass concrete. (c) Counterbalancing moment provided by tension pile.

coincident with the line of action of the combined loads in order to obtain uniform ground pressures.

7.2.4 Basements

A basement foundation is really a specific example of the raft foundation, the floor of the basement constituting the raft surface. Although the ground pressures resulting from this form of construction are usually low, the large span of the base slab produces high bending moments in the base, which thus requires a thick slab. A base slab of uniform thickness over the full width of the basement, known as a rigid base, has these characteristics.

An alternative solution is that known as the flexible base, in which the base slab is of nominal thickness over the central portion. This produces advantages in that the bending moments in the slab are reduced because of the lesser stiffness, but leads to higher bearing stresses under the thickened portion of the slab. Obviously the choice of construction depends on the soil properties of the site.

In addition to the various factors that require attention for all foundation designs, the use of basement foundations frequently creates further problems in that the increased depth to which the foundation is to be placed may put the basement below the water table level, with the result that there may be difficulty with the watertightness of the structure, and that when no loads are on the foundation, i.e. at the time of construction when the basement is completed before the structure itself, the basement may float. This last possibility should, of course, also be checked when the structure is completed, but it is unlikely that the problem will exist at that stage. The various methods that may be used to prevent flotation include temporary flooding of the basement, addition of extra load, and pumping of the site to reduce the water table level. The basement walls must be designed to resist, in addition to the upward hydrostatic pressures, the horizontal pressures that arise both from soil retention and from the head of water retained. Since there may be some seasonal variation in the water table, it is important to ensure that the worst condition is considered.

7.2.5 Piers and piles

Piers and piles rely on frictional forces on the side of the pile, on the transmission of the foundation loads to a more solid stratum of soil, or on a combination of these two effects. Discussion of such foundations is more suited to a specialist text on soil mechanics, as their influence on the subsoil performance is rather more complex than that of other foundation forms. Their use will not therefore be considered here.

7.3   Design examples

The design of foundations follows the general principles of reinforced concrete design as considered in the previous chapters. The detail design is as for beams and slabs, particularly with regard to the bending moments and the shear forces in the foundation. The design procedure is as described earlier (notably in Chapters 3 and 4), and is illustrated here by means of worked examples.

7.3.1   Isolated column footing (Example 7.1)

Design a suitable foundation to carry a 0·5 m square column, if the working load on the column is 2500 kN, and the maximum allowable ground bearing pressure is 0·2 N/mm^2 (approximately 2 tonf/ft^2). Use concrete of Grade 30 specification and mild steel reinforcement, and assume a load factor of 1·6.

The allowable bearing pressure given in this example would typically be applicable to a medium chalk or a hard compact sand. Although the dead and live load elements of the working or service load would usually have differing load factors, in this example the same factor of 1·6 is applied to both, for convenience.

In calculating the base area from the allowable ground bearing pressure, allowance should be made for the self-weight of the base itself. As an initial approximation, an allowance of 5 per cent of the total axial load provides a reasonable value for the base self-weight; if the column has an appreciable bending moment, this allowance should be increased. In this case there is no bending moment, so a figure of 5 per cent is used, giving a self-weight allowance of 125 kN and making the total load on the ground beneath the base 2625 kN. Assuming that the pressure beneath the base is the maximum allowable, the minimum base area is

$$\frac{2625 \times 10^3}{0\cdot2} = 13 \times 10^6 \text{ mm}^2$$

which, for a square base, gives a minimum size of 3622 mm square. It is as well to make the base slightly larger, however, so try a base 3750 mm square. The ground pressure beneath the base is then 0·187 N/mm^2.

The thickness of the base is determined from moment and shear considerations. The minimum thickness for bending is easily established: the maximum bending moment that a section can support is given by eq. (3.6), which may also be used to determine the minimum section thickness for a given bending moment. The critical section for bending in the base is taken as the face of the column (section $X-X$, Fig. 7.5) but,

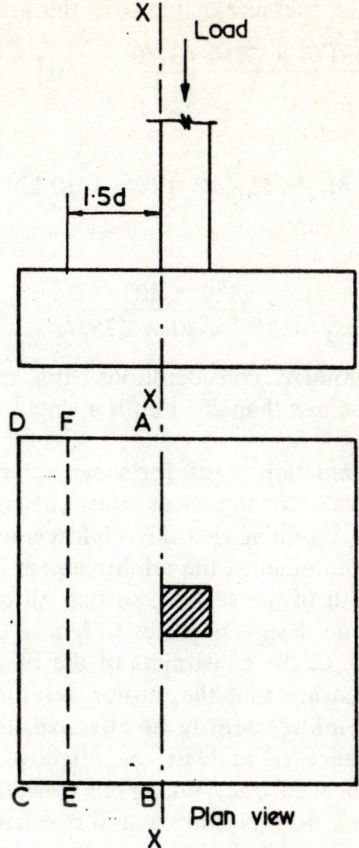

Fig. 7.5 Critical section for bending in an isolated footing.

as far as the bending moments in the foundation are concerned, the self-weight of the footing is acting in opposition to the ground pressure beneath the foundation, so that the moment in the base is caused by the ground pressure beneath the footing acting over the area *ABCD*, less the self-weight of the foundation itself. A similar situation exists in evaluation of the shear forces acting on the foundation, so that the bending and shear in the base are effectively caused by the ground pressures beneath the base that are due to the column loads only, and the self-weight of the base may be ignored in these calculations.

The pressure beneath the base due to the column loads only

$$= \frac{2500 \times 10^3}{3750^2} = 0{\cdot}178 \text{ N/mm}^2$$

M_{XX} is caused by the pressure acting over the area $ABCD$:

$$M_{XX} = \frac{0 \cdot 178 \times 3750}{2} \left(\frac{3750}{2} - 250\right)^2 = 881 \text{ kN m}$$

Therefore

$$M_u = M_{XX} \times 1 \cdot 6 = 1410 \text{ kN m}$$

and, from eq. (3.6),

$$d = \sqrt{\left(\frac{1410 \times 10^6}{0 \cdot 155 \times 30 \times 3750}\right)} = 284 \text{ mm}$$

Therefore, from moment considerations, the effective depth of the foundation cannot be less than 284 mm if a singly reinforced section is to be suitable.

The minimum foundation depth for shear behaviour is not so readily assessed, since the value of the shear stress that is used depends on the reinforcement area. Assuming that the reinforcement area is governed by bending-moment requirements, the reinforcement area therefore depends on the effective depth of the section, so that all of the pertinent factors are interrelated. Some designers prefer to ignore the effect of reinforcement, since, because of the constraints of the base dimensions, it is not always possible to ensure that the tension reinforcement is effective in shear (for tension reinforcement to be effective, the reinforcement must continue for a distance of at least the effective depth of the section beyond the critical shear section); taking this view, the maximum allowable shear stress becomes $0 \cdot 34 \text{ N/mm}^2$ and the calculation is considerably simplified, allowing direct calculation of the minimum effective depth from shear considerations.

As with slabs, there are two possible shear conditions to be considered: (a) the shear across the full width of the section; (b) the punching shear from the concentrated column loads. Of these, the former is the simpler to calculate, and is therefore better suited to preliminary depth estimation purposes, although both conditions should be checked.

By treating the foundation as a slab, the critical section for shear across the full width of the base may be taken as occurring at a distance of $1 \cdot 5 \times$ effective depth from the column face. Ignoring the tensile reinforcement that will be provided for the bending moments in the base (and which has yet to be calculated), the ultimate shear stress in the section is $0 \cdot 34 \text{ N/mm}^2$ (Table 4.2). The load causing shear is due to the pressures beneath the base that act over the area $EFCD$ (Fig. 7.5). These may be assumed to be due to the column load only, and give an ultimate shear force on section EF of $1 \cdot 6 \times 3750 \times 0 \cdot 178 \times FD$ N.

$$FD = \frac{3750}{2} - 250 - 1 \cdot 5d$$

so that the shear stress

$$v = \frac{3750 \times 0 \cdot 178 \times 1 \cdot 6 \times (1625 - 1 \cdot 5d)}{3750 \times d} \text{ N/mm}^2$$

This has a maximum value of $0 \cdot 34$ N/mm^2, so that $d > 603$ mm (say 610 mm) and, allowing for cover to the reinforcement, the actual depth is about 670 mm.

A check on the punching shear stress that would be obtained for an effective depth of 610 mm gives a punching shear stress of $0 \cdot 45$ N/mm^2, showing that, since the maximum allowable punching shear stress is exceeded, some reinforcement is required to enable the punching shear to be carried. From Table 4.2 the reinforcement ratio giving a punching shear stress of $0 \cdot 45$ N/mm^2 is approximately $0 \cdot 38$, so that the reinforcement area required is 8692 mm^2, whereas from bending moment considerations, for an effective depth of 610 mm, moments about the line of action of the tensile force give

$$M_\mathrm{u} = 0 \cdot 4 f_\mathrm{cu} bx(d - 0 \cdot 45x)$$

so that

$$x \simeq 53 \text{ mm}$$

From eq. (3.3), $A_\mathrm{s} = 10\,971$ mm^2, showing that the ultimate punching shear stress on a foundation of 610 mm effective thickness will not be exceeded if the reinforcement area is taken into account.

The design has therefore become a 3750 mm square foundation, having an effective depth of 610 mm and with a reinforcement area of $10\,971$ mm^2 required. This may be obtained from nine 40 mm diameter bars ($A_\mathrm{s} = 11\,309$ mm^2), but this would result in a bar spacing that is greater than that advisable (Chapter 8). Hence use fifteen 32 mm diameter bars ($A_\mathrm{s} = 12\,063$ mm^2), which, allowing for cover at the edges of the slab, gives a spacing of approximately 260 mm.

Since the footing is square and under uniform pressure, reinforcement is required in both directions, and the area of the transverse reinforcement is potentially the same as that calculated above. The effective depth obviously cannot be the same in both directions, however, and must be either less than or greater than the previous value by the diameter of the reinforcing bars. Assuming that d is reduced, the area of the reinforcement required is given by $A_\mathrm{s} = M_\mathrm{u}/(0 \cdot 87 f_y z)$. This produces an A_s value of $11\,062$ mm^2, which is less than that provided by fifteen 32 mm diameter bars, so that the same reinforcement at the same spacing may be used in both directions. A check on the shear values using the reduced effective depth shows that neither shear nor punching shear ultimate values are exceeded.

Other factors which may affect the design, and which have been so far ignored, are the bond stresses in the reinforcement. These are dealt with in more detail in Chapter 8 and the various relevant equations are simply produced here with no discussion as to their validity or origin.

Anchorage bond   Assuming deformed bars, the design ultimate anchorage bond stress is $2 \cdot 19$ N/mm^2 (see Chapter 8). The anchorage length required is therefore

$$\frac{0 \cdot 87 \times 250 \times 32}{4 \times 2 \cdot 19} = 794 \text{ mm}$$

from the point of maximum stress in the reinforcement. This shows that some of the bars could in fact be curtailed before the end of the section, but this would be of doubtful economy, and simplicity of construction suggests that the preferable solution is to continue all of the reinforcement across the full section width.

The final check is to ensure that an adequate allowance has been made for the self-weight of the base. The actual self-weight of the base is

$$\frac{3750^2 \times 650 \times 2400 \times 9 \cdot 81}{10^9} = 215 \text{ kN}$$

which is somewhat higher than the allowance made; the bearing pressure is

$$\frac{(215 + 2500) \times 10^3}{3750^2} = 0 \cdot 193 \text{ N/mm}^2$$

which is less than the permissible and is therefore satisfactory.

Although the above example is for the simplest possible case (that of an isolated footing under an axial load only), the procedure to be followed where the base is subjected to a bending moment is virtually identical, except that the pressures under the base are not uniform, and care must be taken to design to the worst condition. For the bending moment and shear across the full width of the base this is comparatively simple, but the assessment of punching shear is not so easy because of the non-uniform pressure-distribution. A safe procedure is to assume that the value of the punching shear stress is the same all round the critical perimeter, and that this value is that obtained at the worst edge of the perimeter. Although this is obviously conservative, it does provide a safe solution and may therefore be used with confidence in the design.

7.3.2   Combined column footing *(Example 7.2)*

The second example is rather more complex, and concerns a foundation carrying unequal loads from two columns (Fig. 7.6). The allowable ground bearing stress is taken as $0 \cdot 25$ N/mm^2, the concrete is assumed to be of

Grade 30, and the reinforcement to be mild steel with $f_y = 250 \text{ N/mm}^2$. A load factor of 1·6 is applied to all loads.

Taking moments about some arbitrary point A shows that the centroid of the column loads is at a point 1·35 m from the line of action of the 1700 kN load. Unless the centroid of the base is aligned with this point, the pressure distribution beneath the base will be non-linear, and the maximum pressure will be given by

$$\text{pressure} = \frac{\text{load}}{\text{area}} \pm \frac{\text{bending moment}}{\text{foundation section modulus}}$$

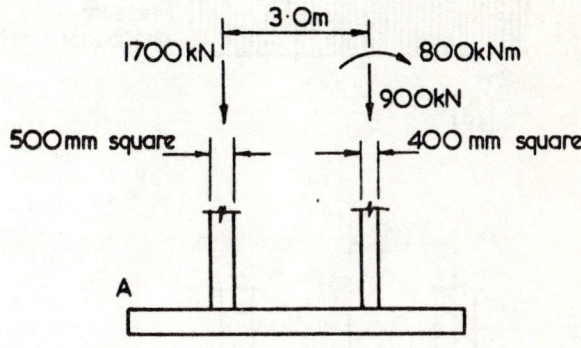

Fig. 7.6

where the bending moment is the load × its eccentricity from the base centroid.

Try a base 2·5 m wide × L m long. As far as the ground pressures are concerned, the column loads may be equated to a single load of intensity equal to the sum of the two individual loads, and acting through the load centroid. Note that in calculating the size of the foundation required to satisfy the ground pressure requirements, no load factor is applied.

Two primary variations in the basic foundation design are possible (Fig. 7.7): either the base may be placed equally about the load centroid, so that the centroids of both load and base coincide (thus giving a uniform distribution of pressure beneath the foundation), or else the base may be placed so that the base centroid coincides with a point midway between the columns. In this case a non-uniform pressure distribution beneath the base will be obtained.

Consider the situation shown in Fig. 7.7(a) where uniform pressure is obtained beneath the foundation. Allowing 10 per cent of the total column load for the self-weight of the foundation, and assuming that a

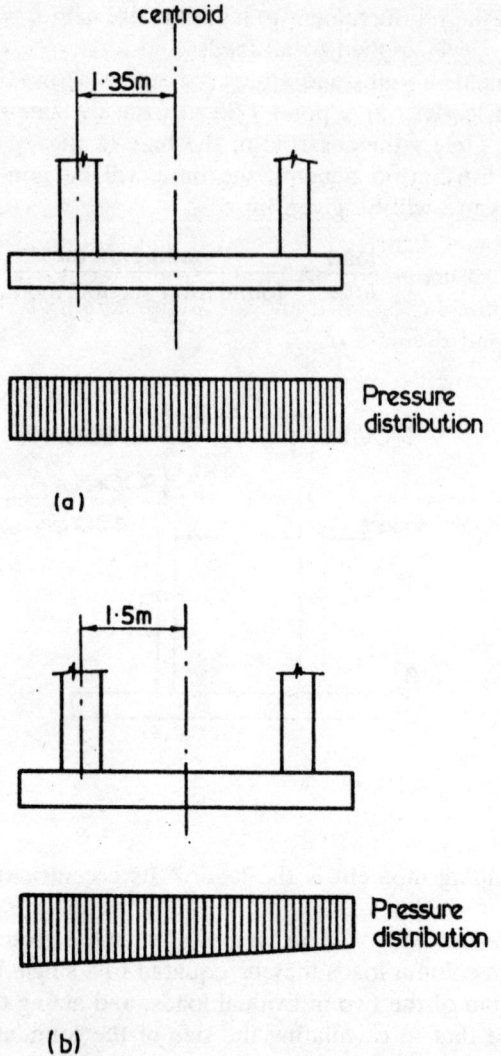

Fig. 7.7 (a) Base and load centroids coincident. Uniform ground pressures.
(b) Base evenly spaced about the columns. Non-uniform ground pressures.

suitable base width is 2·5 m, the length of the foundation required in order that the allowable ground pressures are not exceeded is

$$\frac{2600 \times 10^3 \times 1·1}{2500 \times 0·25} = 4·58 \text{ m}$$

If the base centroid is to be at the point shown in Fig. 7.7(a), the base length must be at least $2(1\cdot65 + 0\cdot2 + \text{thickness})$ since the base should continue past the concentrated load (in this case the column) for a distance approximately equal to the base thickness in order to allow the stress concentration due to the column load to disperse. Assuming a foundation thickness of $0\cdot75$ m, the minimum base length must therefore be $2(1\cdot65 + 0\cdot2 + 0\cdot75) = 5\cdot2$ m. This is greater than the minimum required for ground bearing purposes, so that the foundation width may be reduced. The minimum practicable width would be approximately $0\cdot5 + 2 \times$ thickness (i.e. 2 m), and for a base length of $5\cdot2$ m this would produce a ground pressure of

$$\frac{2600 \times 1\cdot1 \times 10^3}{5\cdot2 \times 2\cdot0 \times 10^6} = 0\cdot275 \text{ N/mm}^2$$

which is above the maximum allowable, and is therefore unacceptable. The size of the foundation must therefore be greater than that above in order to reduce the ground bearing pressures; increasing the foundation to $5\cdot25 \times 2\cdot25$ gives a ground pressure of $0\cdot242$ N/mm^2, which is less than the maximum allowable and is thus satisfactory.

The self-weight of the foundation should now be checked to ensure that an adequate allowance has been made. Assuming the density of reinforced concrete to be 2400 kg/m^3, the self-weight of the foundation is

$$2\cdot25 \times 5\cdot25 \times 0\cdot75 \times 2400 \times 9\cdot81 = 208 \text{ kN}$$

The allowance of $0\cdot1 \times 2600$ kN that has been made is therefore sufficient and, assuming the load and base centroids to be coincident, a suitable foundation size is $5\cdot25$ m $\times 2\cdot25$ m.

Now consider the alternative solution. The base is positioned so that the centroid is at a point midway between the columns, so that the eccentricity e of the load centroid about the base centroid is $0\cdot15$ m. Assuming the base to be of length L and width $2\cdot25$ m, the maximum pressure beneath the base $= (P/A) + (Pey/I)$. Allowing 10 per cent of the total column loads for the foundation self-weight, maximum ground pressure is

$$\frac{2600 \times 10^3 \times 1\cdot1}{2250L} + \frac{2600 \times 10^3 \times 0\cdot15 \times 10^3 \times 6}{2250L^2} \text{ N/mm}^2$$

Note that the self-weight allowance is applied only to the first part of this expression, since the foundation itself produces uniform pressures only.

Putting the maximum pressure equal to the maximum allowable ground pressure gives $L = 5\cdot8$ m, so that a foundation size of $5\cdot8$ m $\times 2\cdot25$ m results.

The most economical solution will be that which ensures a uniform pressure-distribution, since the area of the base is smaller than in the other case considered. If the positions and magnitudes of the loads give a uniform ground pressure beneath the base that is close to the maximum allowable, the solution in which the centroids of both the base and load are coincident always proves more economical, since a non-uniform pressure-distribution involves some part of the base being subjected to ground pressures that are usually well below the maximum permissible.

As with the isolated footing, the shearing forces and bending moments in the base are caused by the pressures beneath the base that result from the column loads only. These pressures must be multiplied by the load factor to allow the ultimate shear and bending moments to be calculated. The ground pressure due to the column loads only is 0.22 N/mm², so that the effective pressure to be considered when calculating the shear force and bending moments is $0.22 \times 1.6 = 0.352$ N/mm².

The shear force and bending moment diagrams are shown in Fig. 7.8.

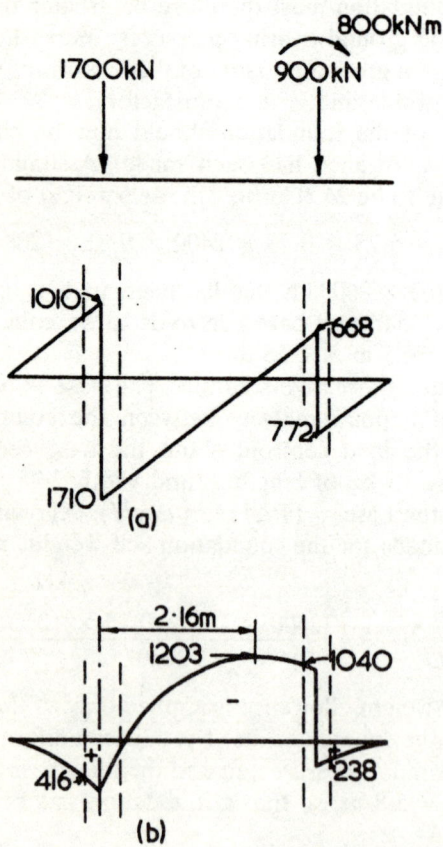

Fig. 7.8 (a) Shear force diagram (kN). (b) Bending moment diagram (kN m).

From Fig. 7.8(b), the maximum and minimum bending moments are −1203 kN m at a point 2·16 m from the centre line of the larger column, and +416 kN m at a section along the larger column face. From eq. (3.6), the minimum effective base depth required to carry the larger of these is 340 mm, giving an actual depth (after allowing for cover) of approximately 400 mm. Bending-moment considerations therefore suggest that the assumed thickness may be reduced. Suppose that the actual thickness is reduced to 500 mm so that a trial effective thickness of 440 mm may now be assumed.

Shear. Since the foundation is to act as a slab, both punching shear and direct shear must be considered. The critical section for shear acting over the full width of the section is at a section 1·5 × effective slab depth away from the face of the column. Assuming an effective depth of 440 mm, therefore, the critical shear force will be 989 kN, acting at a section 660 mm from the face of the larger column towards the smaller column (section $X-X$, Fig. 7.9). This will produce a shear stress at the critical section of $(989 \times 10^3)/(2250 \times 440) = 1·0$ N/mm^2.

From Chapter 4, the maximum value of shear stress that is permitted in a Grade 30 concrete slab is 4·38 N/mm^2. This is greater than the actual maximum shear stress, so diagonal compression failure will not occur. The ultimate design shear stress for Grade 30 concrete is dependent upon the longitudinal reinforcement ratio, but has a maximum value of 0·97 N/mm^2, so that shear reinforcement is necessary in order to carry the shear stress above this value.

The placing of shear reinforcement in a foundation of this size involves a considerable amount of steel because of the width of the foundation. In most situations the preferred solution is to thicken the foundation in order to reduce the shear stress acting over the section, but other considerations, such as cost of excavation and the suitability of the site to

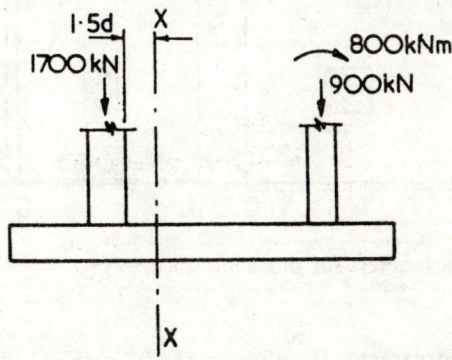

Fig. 7.9 Critical section for shear.

accept deeper foundations, may mean that shear reinforcement is the cheapest solution.

Assuming that there are no restrictions to foundation depth, increase the actual depth to 650 mm. This produces an effective depth of about 590 mm. The maximum shear stress is now 0·61 N/mm², which is acceptable provided that a tensile reinforcement ratio of about 0·75 is obtained. At this stage, however, the reinforcement ratio has not been calculated, so that the exact ultimate shear stress cannot be compared with the value of the critical shear given by the above calculation.

Punching shear. The critical section for punching shear acts on a perimeter 1·5 × the effective slab depth from the concentrated load. In this case the concentrated load is provided by the column, so that the critical perimeter for each column becomes as shown in Fig. 7.10. Part of the critical perimeter for punching shear for both columns falls outside the foundation, so that the critical sections are *AB* and *CD* for the larger column and *EFGH* for the smaller. Section *CD* is obviously the more critical of these, since the shear force is appreciably greater at this section; but this is virtually the same section as that considered as the critical section for shear across the complete width of the foundation, so that in this instance, since the direct shear conditions have been satisfied, no further check of the punching shear across the foundation is required.

The preliminary sizing of the foundation is therefore complete, with the initial checks showing that a base 650 mm deep × 5250 mm long × 2250 mm wide is suitable. The detail design of the reinforcement areas and the checks on the ultimate shear stress must now be carried out.

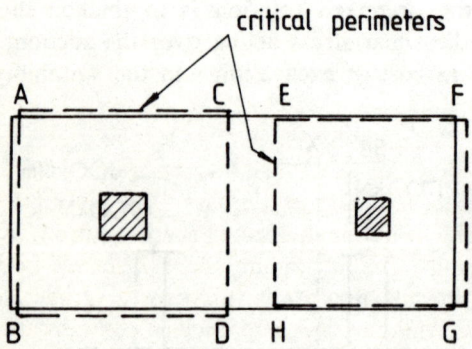

Fig. 7.10 Critical perimeters for punching shear.

Longitudinal reinforcement. Reinforcement is required in the top face of the foundation to carry a maximum bending moment of 1203 kN m.

Reinforcement in this face is only required over the area of negative bending moment (Fig. 7.8(b)), with some extra length added for bond and anchorage lengths. Since the bending moment varies considerably over this area, so the area of reinforcement required will also show considerable variation.

The length of foundation needing reinforcement is such, however, that in practice a constant reinforcement area will be provided over the whole zone of negative bending. The bottom face of the foundation theoretically requires reinforcement only over the area of positive bending moment, but it is unlikely that stopping the bottom face reinforcement will be economic, except perhaps for some reduction in area over the portion of the foundation under the 400 mm square column.

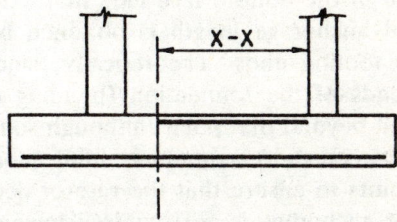

Fig. 7.11 General arrangement of reinforcement.

The general arrangement of the longitudinal reinforcement is as shown in Fig. 7.11. The foundation may be considered as being doubly reinforced over the central portion $X-X$, and singly reinforced over the remainder.

The area of reinforcement required is obtained from the general procedure of Chapter 3, which shows that for a singly reinforced section having an effective depth of 590 mm and a width of 2250 mm, a bending moment of 416 kN m requires an area of longitudinal reinforcement of 3360 mm^2, and a bending moment of 238 kN m an area of 1900 mm^2. For simplicity of construction these areas may be obtained from nineteen 16 mm diameter bars reducing to ten bars. This gives areas of 3821 mm^2 and 2011 mm^2, which, although greater than required, provide a simple reinforcement detail. The maximum clear distance between the bars will be less than the maximum allowable for a Grade 30 concrete (see Chapter 8), so that the bar sizes and spacings chosen are suitable.

The area of reinforcement required in the top of the foundation may also be assessed by means of the procedures of Chapter 3. Treating the section as being doubly reinforced over the portion of negative bending moment gives a required reinforcement area of 9943 mm^2, which may be obtained from thirteen 32 mm diameter bars (area 10455 mm^2).

The tensile reinforcement ratio is therefore 0·787, for which the ultimate shear stress value is 0.62 N/mm^2. This is in excess of the maximum shear stress of 0.61 N/mm^2, so that the shear conditions are satisfied.

Anchorage. From BS 8110, assuming that plain bars are used, the full anchorage length required is $36 \times$ bar diameter for bars in tension, and $29 \times$ bar diameter for bars in compression. Assuming all of the bars to be in tension, anchorage lengths of 576 mm (say 600 mm) for the 16 mm bars and 1152 mm (say 1200) for the 32 mm diameter bars are required. No check of the anchorage length for bars in compression is needed as there will be no curtailment of the compression reinforcement which is carried through.

The above anchorage lengths assume that the bars are fully stressed, and are therefore to be measured from the points of maximum bending moment. In the case of the bottom face bars that extend to the ends of the footing, the full anchorage length is obtained by carrying the reinforcement to the footing ends. Theoretically, since no stress exists in the bar at the ends of the foundation (bending moment zero), no anchorage is required beyond that point, although some check should be made at sections between the maximum bending moment and the zero bending moment points to ensure that the rate of decay of the bending moment is such that anchorage is not required beyond the point where bending moment becomes zero. Some designers prefer to provide the full anchorage length at the end of a bar, whether it is required or not—a practice which may be uneconomic in terms of materials, but saves time at the design stage.

The full anchorage length for the top reinforcement (assuming that stress in the reinforcement is pro-rata with bending moment) for a bending moment of 1040 kN m is 935 mm. Although this is less than the distance from the inside face of the smaller column (where the bending moment is 1040 kN m) to the edge of the foundation, so that in theory a straight bar would have sufficient anchorage length, the distance from the column face to the foundation edge is such that good practice would be to provide an additional anchorage in the form of a hook or bend in the reinforcement, so that the reinforcement detail becomes as shown in Fig. 7.12.

Transverse reinforcement. In addition to the longitudinal reinforcement, transverse reinforcement is required to carry the bending moment acting across the section. Taken at the plane of the smaller column face (as this is the critical section), the bending moment is

$$\frac{925^2}{2} \times 0.352 \times 5250 = 791 \text{ kN m}$$

Assuming a singly reinforced section, $d \sim 570$ mm, so that from the procedures of Chapter 3 the required reinforcement area is 1191 mm^2/m.

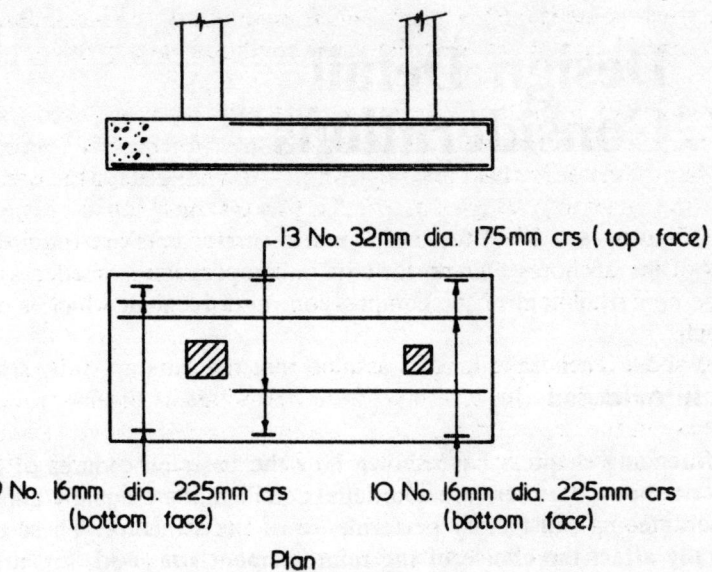

−13 No. 32mm dia. 175mm crs (top face)

9 No. 16mm dia. 225mm crs
(bottom face)

10 No. 16mm dia. 225mm crs
(bottom face)

Plan

Fig. 7.12 Longitudinal reinforcement detail.

This is less than the minimum reinforcement area that should be provided (Chapter 8), which is 0·24 per cent *bh*, so that the reinforcement area specified should be at least the minimum area of 1800 mm^2/m. Bars of 20 mm diameter at 175 mm spacing provide sufficient area and comply with the spacing requirements. A check on the shear acting on this section shows that the shear stress in the foundation is below the ultimate allowable for the reinforcement ratio provided, and consideration of the anchorage bond indicates that straight bars give sufficient anchorage length. The design may therefore be considered to be complete.

8 Design Detail Considerations

8.1 Introduction

The foregoing chapters have shown how the basic procedures of design are carried out, but there are additional details that are equally important in guaranteeing satisfactory performance of the structure. These details generally affect the choice of the reinforcement size used, for although the calculations of the previous chapters allow the reinforcement area to be calculated, choice of bar size will depend on a number of factors, not least of which may be economy. As a general rule, the more bars included the more expensive the structure, for the unit costs of bending and fixing reinforcement are largely independent of bar size, so that a greater number of bars leads to a greater cost. The basic cost of reinforcement is based on a 25 mm diameter, 12 m long bar. Bars of smaller diameter carry a price premium, as do lengths shorter than 12 m, whereas all larger diameter bars are supplied pro rata with the 25 mm dia size. Reinforcement lengths greater than 12 m are also more expensive than the basic lengths; extremely long lengths of reinforcement have the additional disadvantage that they cause transportation problems both on and off the site.

8.2 Bond and anchorage

Bond is the longitudinal force between concrete and reinforcement. It is achieved by shrinkage of the concrete during the setting action, which grips the reinforcement, so that bond may be considered to be due to a combination of adhesion and friction between the concrete and the reinforcement. The adhesion element is low, and breaks down under quite small amounts of slip between the concrete and the reinforcement; frictional and bearing effects then supply the remaining bond strength. If the reinforcement yields, the reduction in diameter reduces the grip of the concrete on the bar and allows the bar to pull out of the concrete at a reduced load. For this reason, care should be taken to ensure that plain bars in particular are adequately anchored in the concrete.

Deformed bars have a higher bond strength than do plain bars, since the deformations increase the area of contact between the concrete and the reinforcement; in addition, the bond strength of deformed bars having projections on the surface of the bar is increased by the bearing between the projection and the surrounding concrete. Deformed bars also have the advantage that they are less likely to pull out of the concrete.

For design purposes, checking of the bond between the concrete and the reinforcement is carried out for two conditions. Local bond (sometimes referred to as flexural bond) is the bond that results from the rate of change of the force in the reinforcement at the section in question, whereas anchorage bond is that bond necessary to ensure that the reinforcement can develop the required force. Consideration of the anchorage bond therefore allows the designer to calculate the length of anchorage required in order to obtain the necessary stress in the reinforcement.

Local bond. From the definitions of local bond, the local bond force = $(dM)/(z\,dx)$ since M/z is the force in the reinforcement, and $(dM)/(z\,dx)$ is therefore the rate of change of force in the reinforcement. But dM/dx = shear force V, so that local bond force = V/z per unit length.

The area on which the bond operates is the sum of the surface areas of the reinforcement at that section, hence

$$\text{local bond stress} = \frac{V}{z\Sigma u_s} = f_{bs} \tag{8.1}$$

where Σu_s is the sum of the effective perimeters of the reinforcing bars at the section in question.

Although previous codes of practice considered local bond as a separate calculation, it is now recommended that as long as suitable anchorage is provided to the bar such that the force required in the reinforcement may be adequately developed, then local bond stress may be ignored.

Anchorage bond. Anchorage bond is defined as

$$\frac{\text{force in the bar}}{\text{anchorage area}}$$

so that the design anchorage bond stress is

$$\frac{\text{force in the bar}}{\text{anchorage length} \times \text{effective perimeter of the bar}}$$

Although research indicates that the development of the anchorage bond force varies along the bar, for design purposes the anchorage bond is taken as being constant along the anchorage length. In order to prevent anchorage bond failure, the stress calculated by the above expression must be restricted to a value that is below the design ultimate anchorage bond stress. This is obtained from the equation

$$f_{bu} = \beta\sqrt{f_{cu}} \qquad\qquad (8.2)$$

where β is a coefficient that depends on the type of bar and the direction of stress in the bar. Values of β are shown in Table 8.1. Hence, for a bar of diameter ϕ, with a stress in the reinforcement f_s and having a design ultimate anchorage bond stress of f_{bu},

$$\text{anchorage length required} \not< \frac{f_s\phi}{4f_{bu}}$$

so that, for a 16 mm dia bar under a stress of 217 N/mm^2 and having a design ultimate anchorage bond stress of 1·5 N/mm^2, the anchorage length required is at least 580 mm.

Bar type	β	
	Bars in tension	Bars in compression
Plain bars	0·28	0·35
Type 1: deformed bars	0·40	0·50
Type 2: deformed bars	0·50	0·63
Fabric (see 3.12.8.5)	0·65	0·81

(Reproduced courtesy of BSI. Source: BS 8110 Part 1, Table 3.28)

It will frequently be found that practical considerations make development of the full anchorage length difficult to achieve for straight bars due to the proximity of the ends of the section or some other constraint. In such cases it is quite permissible to use hooks or bends in the ends of the bars. These fulfil two functions: firstly they allow the full anchorage length to be developed, and secondly some additional component of anchorage is obtained by virtue of the bearing of the inside of the bend on the concrete. This latter component has the effect of increasing the effective anchorage length of a hook or bend, and although the effective anchorage provided by a hook or a bend is covered in detail in the appropriate BS 8110 clauses, it may be summarised as being that the effective anchorage of a hook or bend (measured from the start of the bend to a point 4 × bar diameter beyond the end of the bend) may be taken as the lesser of 24 × bar diameter or:

(i) for a hook, 8 × internal radius of the hook;
(ii) for a 90° bend, 4 × internal radius of the bend.

If the bar is assumed to be unstressed at a point 4 × bar diameter past the end of the bend, there is no need to check the bearing stress on the inside of the bend or hook. For any other situation, the bearing stress inside the bend should be checked, and BS 8110 gives details of the procedure to be followed.

8.3 Lapping of reinforcement

The length of reinforcement required will frequently be such that reinforcing bars have to be joined. This may be achieved by welding or by means of some mechanical connection, but traditionally the most common form of transferring the stress from one bar to another is by overlapping the bars so that the stress transfer is effected through the anchorage of the bars in the concrete. The lap length that should be provided is dependent upon a number of factors that are detailed in BS 8110, but the recommendations may be broadly summarised. The length of lap provided should be at least equal to the anchorage length of the smaller of the bars being lapped, subject to the provision that the minimum lap length should not be less than 15 × the bar size, or 300 mm.

8.4 Curtailment of reinforcement

The amount of reinforcement that is placed in a member is calculated for the maximum bending moment in the section. Unless the section is under a uniform bending moment it is uneconomical to provide the same reinforcement throughout, and advantage may be taken of a reduction in bending moment to reduce the reinforcement area provided. However, although it may be uneconomical to provide a constant area of reinforcement throughout a section, it may be equally uneconomical to be constantly changing the reinforcement size and spacing, so that attention must be paid not only to the design requirements, but also to the practical disadvantages that may result from changing the amount of reinforcement provided.

Reduction of the reinforcement area is best achieved by stopping or curtailing bars rather than by changing bar size. Careful attention must be given both to ensuring that sufficient anchorage length is provided to the curtailed bars to allow the required stress to be developed in the bar and to the spacing of the remaining bars (Section 8.5). The reinforcement is curtailed at a point known as the 'theoretical cut-off point' (TCP). At this point the moment of resistance of the section excluding the curtailed bars is equal to or greater than the applied bending moment; the bar should, however, be extended beyond the point at which it is no longer needed (the TCP) for some distance to allow for any inaccuracies that may occur in the design or in the construction. A suitable distance is recommended to be the greater of either the effective depth of the section or 12 × bar diameter.

In addition, for curtailment of reinforcement in a tension zone, one of three extra conditions must be satisfied:

1. The bars must extend beyond the TCP for the full anchorage length.
2. The shear capacity of the section where the reinforcement is curtailed must be greater than twice the applied shear force at that section.
3. The reinforcement continuing past the section where the bars are curtailed must provide at least double the area required to resist the moment at that section.

These three conditions are included to control the size of cracks that may occur at the cut-off points, and to ensure an adequate reserve of shear strength. However, even if Condition 1 above is not directly checked (since the code requirement is that only one of the three conditions must be satisfied), the bond conditions must be examined to ensure that the bond stresses are not exceeded.

Simplified rules for the curtailment of reinforcement in beams and slabs are presented in BS 8110. These are frequently easier to apply to the practical design situation, but may not produce quite such an economical detail as the conditions outlined above.

8.5 Spacing of reinforcement

An important factor in determining the size of the bars to be used is the spacing of bars that will result. The distance between bars must lie between certain limits, the lower limit being a function of the aggregate size being used, while the upper limit is necessary to avoid the development of excessively wide cracks in the concrete.

The exact minimum spacing depends on the arrangement of the reinforcement (details are given in BS 8110); for individual bars the horizontal distance between bars should not be less than the maximum aggregate size plus 5 mm, or the bar diameter, whichever is the greater. This is primarily to ensure that full compaction of the concrete can be achieved, particularly around the reinforcement. The maximum spacing between the bars is limited to restrict the widths of cracks that occur. Excessively wide cracks are not only unsightly and to the layman possibly worrying, but may permit moisture to reach the reinforcement and cause corrosion. Recommendations as to the maximum bar spacing are therefore also given in BS 8110, but these are rather lengthy and will not be reproduced here.

8.6   Cover to the reinforcement

The spacing of the reinforcement is mainly of interest at the detailing stage, and does not give great concern to the designer. This is not the case, however, with the cover that is provided, since the effective depth of the section is dependent upon the actual depth and the cover. The worked examples of the preceding chapters have presumed cover to be taken to the main reinforcement. This has been done for arithmetical convenience only; the actual cover given should be measured from the concrete surface to the outermost fibre of any reinforcement, including stirrups or links.

The need to provide adequate cover to the reinforcement follows from three basic requirements. The first is that the reinforcement should be adequately gripped by the concrete (the bond referred to in Section 8.2). The second and third arise from the need to protect the reinforcement, firstly against corrosion and secondly against fire. The amount of cover that must be provided for these two latter conditions is dependent upon a number of requirements. In particular, the subject of fire resistance

Table 8.2   Nominal cover to reinforcement

| Condition of exposure | Nominal cover (mm) Concrete grade | | | |
	30	35	40	50
Mild: e.g. completely protected against weather or aggressive conditions.	25	20	20	20
Moderate: e.g. sheltered from severe rain and against freezing while saturated with water. Buried concrete in a non-aggressive soil, and concrete continuously under water.	—	35	30	20
Severe: e.g. exposed to driving rain, alternate wetting and drying, occasional freezing, or to heavy condensation.	—	—	40	25
Very severe: e.g. exposed to sea or moorland water, or to de-icing salts. Subject to severe freezing or to corrosive fumes.	—	—	50	30

is extremely complex, but the procedures to be followed are comprehensively covered in BS 8110 which presents details in tabulated form of the cover required for different degrees of fire protection. The degree of corrosion protection that is necessary depends on the exposure conditions and on the basic quality of the concrete, and Table 8.2 summarises the details that are given in BS 8110 concerning the nominal cover that should be provided to ensure a durable concrete. In addition to details of cover, Table 8.2 also gives indication of the minimum grade of concrete that should be used for the various conditions.

8.7 Durability

The completed concrete structure should obviously perform satisfactorily throughout its design life, and in order to achieve this it is necessary to ensure that the structure has adequate strength, not only at the time of design and construction, but also that the strength is maintained with no deterioration. Hence the need for a durable concrete, which will protect the reinforcement and maintain strength. An important consideration in concrete durability is permeability, not only to resist the ingress of water, but also in respect of absorbtion of carbon dioxide from the atmosphere (leading to a condition known as carbonation), and exposure to other chemicals. In order to achieve an adequately low permeability, care must be taken to ensure adequate cement content, correct compaction, and proper hydration of the cement. The latter point requires attention to both the water/cement ratio and to the curing process. Guidance on all of these points is given in the British Standard.

8.8 Limits of reinforcement area

The area of main reinforcement required in a reinforced concrete section is normally determined by the design calculations. The area provided, however, is subject to upper and lower limits, which are applied for largely practical reasons. The upper limit on the area of reinforcement is necessary in order to prevent congestion of the reinforcement which makes full compaction difficult to achieve. In order to avoid this, the recommendations are that for beams neither the area of tension reinforcement nor the area of compression reinforcement should exceed 4 per cent of the gross cross-sectional area of concrete, and that for columns the area of longitudinal reinforcement should not exceed 6 per cent for vertically cast columns, or 8 per cent for horizontally cast columns, once again in terms of the gross cross-sectional area.

The limitation on the minimum amount of reinforcement provided applies to sections that for various reasons are much larger in cross-section than is necessary to carry the applied loads. In such cases, the ultimate moment of resistance of the section may be less than the moment causing cracking so that, if the load becomes sufficient to cause tensile cracking of the concrete, a sudden failure occurs with no warning. This is the extreme of the under-reinforced section, when the reinforcement may be stressed so highly and so quickly that the usual large deformations occur virtually instantly. Failure of this nature may be eliminated by ensuring that a minimum area of reinforcement is provided, which BS 8110 recommends as being 0·24 per cent bh for mild steel reinforcement, and 0·13 per cent bh for high yield reinforcement. These percentages apply only for rectangular beams under flexural conditions, and different values are recommended for other cross-sectional shapes and load situations. Details are given in BS 8110.

In addition to the main reinforcement that is necessary to support the various tensile stresses, some amount of transverse reinforcement is required, particularly in slabs and flanged beams. This is needed in order to control shrinkage, to give adequate shear capacity, and to spread the effects of concentrated loads, but the minimum amount of transverse reinforcement that should be provided depends on the structure. For rectangular sections, including solid slabs, the minimum amount of transverse reinforcement is the same as the minimum amount of main reinforcement detailed above, but for flanged beams the minimum amount (which should be placed near the top surface of the flange) becomes 0·15 per cent $h_f l$, where l is the span of the beam. This applies to both mild steel and to high yield reinforcement.

Appendix

Comparison of the rectangular-parabolic and the linear-rectangular compression stress blocks for concrete

The design stress−strain curve for concrete is shown in Fig. 3.4. Assuming $\gamma_m = 1.5$, the assumed stress-distribution in a reinforced concrete beam under flexure is as shown in Fig. A.1(a) for the rectangular-parabolic compressive stress block, and in Fig. A.1(b) for the linear-rectangular stress block. Figure A.1(c) shows the strain diagram across the section.

The important parameters required from the two assumed stress blocks are the total compressive force, and the product of compressive force × lever arm when taken about the line of action of the tensile forces. These give the area of tension reinforcement required and the ultimate moment of the section.

Considering the rectangular-parabolic case, and assuming that the stresses are acting on a rectangular beam of unit width, the compressive

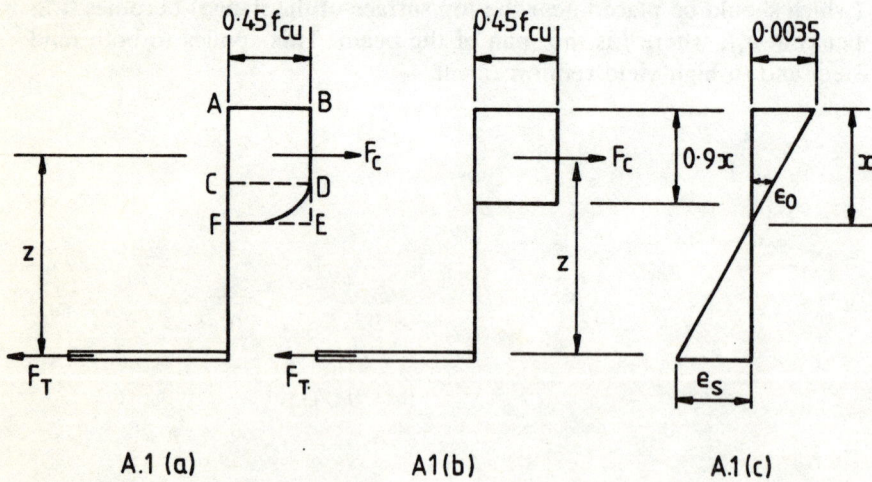

A.1 (a)   A.1 (b)   A.1 (c)

Fig. A.1

force is given by the area of the stress diagram. This is obtained from the area of rectangle $ABCD$ plus $\frac{2}{3}$ area rectangle $CDEF$; but $AC = x - CF$ and, since the strain at CD is ϵ_0, from the strain diagram

$$CF = \frac{\epsilon_0 \times x}{0 \cdot 0035}$$

therefore

$$AC = x \left(1 - \frac{\epsilon_0}{0 \cdot 0035}\right)$$

and the area of the stress diagram is

$$0 \cdot 45 f_{cu} x \left(1 - \frac{\epsilon_0}{0 \cdot 0035}\right) + \tfrac{2}{3} \times 0 \cdot 45 f_{cu} x \left(\frac{\epsilon_0}{0 \cdot 0035}\right)$$

$$= 0 \cdot 45 f_{cu} x \left(1 - \frac{\epsilon_0}{0 \cdot 0035 \times 3}\right)$$

but

$$\epsilon_0 = 2 \cdot 4 \times 10^{-4} \sqrt{\frac{f_{cu}}{\gamma_m}}$$

so that

$$F_c(a) = 0 \cdot 45 f_{cu} x \left(1 - \frac{\sqrt{f_{cu}}}{53 \cdot 6}\right)$$

The lever arm is obtained by considering the line of action of this force. Suppose that the line of action is through a point $\bar{x}$ down from the compression extreme fibre.

Taking moments about AB,

$$F_c(a)\bar{x}(a) = 0 \cdot 45 f_{cu} \frac{AC^2}{2} + \tfrac{2}{3} \times 0 \cdot 45 f_{cu} CF \bar{\bar{x}}$$

where $\bar{\bar{x}}$ is the distance of the centroid of the parabolic area CDF from the compression extreme fibre.

$$\bar{\bar{x}} = AC + \tfrac{3}{8} CF = x \left(1 - \frac{\epsilon_0}{0 \cdot 0035}\right) + \tfrac{3}{8} \times \frac{\epsilon_0 x}{0 \cdot 0035}$$

$$= x \left(1 - \frac{5\epsilon_0}{8 \times 0 \cdot 0035}\right)$$

Therefore

$$F_c(a)\bar{x}(a) = \frac{0 \cdot 45}{2} f_{cu} x^2 \left(1 - \frac{\epsilon_0}{0 \cdot 0035}\right)^2$$

$$+ \tfrac{2}{3} \times 0 \cdot 45 f_{cu} x^2 \frac{\epsilon_0}{0 \cdot 0035} \left(1 - \frac{5\epsilon_0}{8 \times 0 \cdot 0035}\right)$$

but

$$F_c(a) = 0.45 f_{cu} x \left(1 - \frac{\epsilon_0}{0.0035 \times 3}\right)$$

therefore

$$\bar{x}(a) = \frac{x}{2}\left(1 - \frac{\epsilon_0}{0.0035}\right)^2$$
$$+ \frac{2\epsilon_0 x}{3 \times 0.0035}\left(1 - \frac{5\epsilon_0}{8 \times 0.0035}\right)\left(1 - \frac{\epsilon_0}{0.0035 \times 3}\right)$$

where

$$\epsilon_0 = 2.4 \times 10^{-4} \sqrt{\frac{f_{cu}}{\gamma_m}}$$

The lever arm (a) is given by $d - \bar{x}(a)$.

Table A1

Characteristic concrete strength (N/mm^2)	Compressive force (N)		$\bar{x}(a)$	$\bar{x}(b)$	Ultimate design moment (N mm)	
	$F_c(a)$	$F_c(b)$			$M_u(a)$	$M_u(b)$
20	8·23x	8·1x	0·46x	0·45x	6·33d	6·24d
30	12·1x	12·1x	0·45x	0·45x	9·37d	9·36d
40	15·8x	16·2x	0·44x	0·45x	12·32d	12·48d
50	19·47x	20·2x	0·43x	0·45x	15·28d	15·6d

d is the effective depth of the section, and x is the depth from the compressive extreme fibre to the neutral axis.

For the linear-rectangular stress-distribution,

$$F_c(b) = 0.4 f_{cu} x$$

The line of action of this force is $0.45x$ down from the compressive force, so that the lever arm (b) $= (d - 0.45x)$.

A comparison of the results obtained from these two stress-distributions is given in Table A1. This shows that very little difference is obtained in the two values of total compressive force or ultimate design moment, the latter of which is calculated for the limiting condition of $x = d/2$.

Index

DESIGN OF REINFORCED CONCRETE ELEMENTS

SECOND EDITION